Air Base Attacks and Defensive Counters

Historical Lessons and Future Challenges

Alan J. Vick

RAND Project AIR FORCE

Prepared for the United States Air Force
Approved for public release; distribution unlimited

For more information on this publication, visit www.rand.org/t/RR968

Library of Congress Cataloging-in-Publication Data is available for this publication.

ISBN: 978-0-8330-8884-0

Published by the RAND Corporation, Santa Monica, Calif.
© Copyright 2015 RAND Corporation
RAND® is a registered trademark.

Support RAND
Make a tax-deductible charitable contribution at
www.rand.org/giving/contribute

www.rand.org

Preface

In the past decade, U.S. national security policymakers and defense analysts have given increasing attention to the problem of adversary anti-access and area-denial strategies. These strategies are intended to inhibit U.S. political and operational access to and disrupt U.S. operations in key regions. A central feature of adversary anti-access strategies is capabilities designed to disrupt the operation of U.S. and partner-nation forward air bases. Although emerging systems, such as precision standoff systems, present new challenges to air base operations, attacks on air bases are nothing novel, dating back to the opening months of World War I. In the past century, air base attacks have been common in both small and large wars; specialized tactics and systems have been developed for offensive operations against adversary bases and for the defense of friendly bases.

To better understand this policy problem and explore the relative effectiveness of potential solutions, RAND Project AIR FORCE has conducted a series of assessments over recent years exploring historical, political, and operational aspects of the access problem. This report is designed to complement these more-operational and technical analyses—which are generally not available to the general public, media, and academic communities. The objective is to better inform the public debate on this important policy problem. This report is also intended as a reference for those military planners, strategists, war-college students, and other defense professionals who have not yet had an opportunity to delve deeply into the topic. It presents an overview of the problem, describes the history of air base attacks and defensive counters, and broadly explores emerging challenges and options to enhance the future operability of air bases.

This report integrates and extends an analysis that was originally commissioned by 13th Air Force and Headquarters Air Force Operational Planning Policy and Strategy, as well as research conducted in support of the fiscal year 2014 U.S. Air Force Scientific Advisory Board study "Defense of USAF Forward Bases." The research described in this report was conducted within the Strategy and Doctrine Program of RAND Project AIR FORCE.

RAND Project AIR FORCE

RAND Project AIR FORCE (PAF), a division of the RAND Corporation, is the U.S. Air Force's federally funded research and development center for studies and analyses. PAF provides the Air Force with independent analyses of policy alternatives affecting the development, employment, combat readiness, and support of current and future air, space, and cyber forces. Research is conducted in four programs: Force Modernization and Employment; Manpower, Personnel, and Training; Resource Management; and Strategy and Doctrine. The research reported here was prepared under contract FA7014-06-C0001.

Additional information about PAF is available on our website:
http://www.rand.org/paf

Contents

Figures

Tables

Summary

Background

From the first documented attack by a British aircraft against a German airfield in 1914 to the August 2014 capture of a Syrian air base by insurgents from the Islamic State of Iraq and Syria, airfields have been prominent targets in dozens of conflicts on every continent except Antarctica. In the past 100 years, attacking forces have used a diverse array of weapons, including aircraft, missiles, naval guns, artillery, mortars, rockets, satchel charges, and small arms. Attacker objectives have also been diverse, including destruction of aircraft and equipment, inflicting casualties, denial of use, capture of the base, and harassment of defenders.[1] It is noteworthy that, whether the conflict was minor, as was the case in the four-day-long Soccer War between Honduras and El Salvador in 1969, or global, such as World War II, combatants seem determined to attack adversary airfields. Over the course of these conflicts, airmen developed a set of defensive measures that have helped reduce losses and sustain operations, particularly when used in an integrated and mutually supporting fashion. Thus, air bases have long been a place of contestation between attackers and defenders, a battlespace, although few outside of the air base defender community think of friendly air bases in those terms.

Since the end of the Cold War, U.S. dominance in conventional power projection has allowed U.S. air forces to operate from sanctuary, largely free from enemy attack. This led to a reduced emphasis on air base defense measures and the misperception that sanctuary was the normal state of affairs rather than an aberration. The emergence of the long-range, highly accurate, conventional missile (both ballistic and cruise) as a threat to air bases is now widely recognized in the U.S. defense community, and, with that recognition, there is a growing appreciation that this era of sanctuary is coming to an end. Consequently, there is renewed interest in neglected topics, such as base hardening, aircraft dispersal, camouflage, deception, and air base recovery and repair.

This report is intended to provide a reference on air base attack and defense to inform public debate, as well as government deliberations, on what has become known as the anti-access problem, specifically as it applies to air base operations. The report explores the history of air base attacks in the past century and describes a new American way of war that emerged after the fall of the Soviet Union. It then argues that emerging threat systems are disruptive to this way of war and will require new concepts of power projection. Finally, the report identifies five classes

---

[1] These attacker objectives are drawn from Alan J. Vick, *Snakes in the Eagle's Nest: A History of Ground Attacks on Air Bases*, Santa Monica, Calif.: RAND Corporation, MR-553-AF, 1995.

of defensive options that have proven valuable in past conflicts and offers recommendations on how best to win what Norman Franks called "the battle of the airfields."[2]

Findings

- **Air base attacks have been a common feature of both minor and major conflicts in the past century.** Between 1914 and 2014, there were at least 26 conflicts in which air bases were attacked. The conflicts spanned the globe, including Central America, South America, Europe, Africa, the Middle East, Southwest Asia, South Asia, Southeast Asia, Australia, and Northeast Asia. Air bases have often been priority targets during the early phases of conflicts. Prominent examples include Germany's 1941 Operation Barbarossa against the Soviet Union, Japanese attacks on airfields on Oahu and in the Philippines in 1941, Israeli preemptive attacks against Arab airfields during the 1967 Six-Day War, and Indian and Pakistani air attacks at the start of their 1965 conflict. The United States has the most recent experience with airfield attacks, having struck adversary air bases in Iraq, Afghanistan, Serbia, and Libya in operations between 1990 and 2014. In World War II alone, Axis air forces lost more than 18,000 aircraft destroyed *on the ground* by U.S. Army Air Force, Navy, and Marine air attacks. Air forces are not the only threat to air bases; ground forces destroyed many aircraft on the ground during World War II and the Vietnam War. During the latter, the U.S. Air Force (USAF) lost roughly 1,600 aircraft damaged or destroyed on the ground by Vietcong or North Vietnamese rocket or mortar attacks on U.S. bases in South Vietnam.
- **The major components of air base defense first identified in World War I (active defense; camouflage, concealment, and deception (CCD); hardening; dispersal on and off base; and postattack recovery) reflect enduring military principles and offer a sound framework for air base defense planning today.** From the earliest days of air combat, airmen recognized the threat to air bases and took steps to reduce vulnerabilities. For example, during the first successful attack on an airfield (the October 8, 1914, British attack on a German Zeppelin base), German defenders were prepared and employed medium machine guns against the (single) attacking aircraft. Although this failed to stop the attack, the British aircraft was damaged and had to land behind German lines. World War II saw great advances in defensive techniques, including hardened aircraft shelters; major deception efforts; wide dispersal of aircraft on and across bases; air base recovery efforts; and the integration of radar, radio, observers, aircraft, and anti-aircraft artillery into air defense networks. By World War II, air base defense had developed into a field of military study with its own set of guidelines, if not formal doctrine. For example, a portfolio approach to base defense that integrated all options into a defensive scheme was established as a best practice in the early 1940s and remains so in 2014. Thus, although the specific technologies continue to evolve, the basic outlines of sound air base defense are long established. Finally, the history of past airfield battles offers useful lessons for operators, planners, and policymakers today.
- **After the Cold War ended, the United States found that it could operate from rear-area sanctuaries, and from this flowed a new American way of war.** In the waning

---

[2] Norman L. R. Franks, *Battle of the Airfields: Operation Bodenplatte, 1 January, 1945*, London: Grub Street, 1994.

days of the Soviet Union, the United States found itself in a conflict with Iraq, a nation that had demonstrated substantial military capability. In response to the Iraqi invasion of Kuwait, the United States and allies, respecting Iraqi military potential, moved a massive air, ground, and naval force to the region. U.S. planners believed that superior U.S. air forces and ground-based air defenses could keep the Iraqi Air Force at bay and, as a result, based forces to maximize efficiency rather than minimize vulnerability to attack. Airfields were packed with aircraft and ports jammed with equipment and supplies. As it turned out, the Iraqis were unable to threaten these force concentrations in a serious way, and coalition air–ground forces routed the Iraqi Army in a handful of days (after it had been pummeled from the air for several weeks). This experience established a new American way of war that became the template for Operations Allied Force, Enduring Freedom, Iraqi Freedom, and Odyssey Dawn. The new way of war did the following:

- Rapidly deploy large joint forces to forward bases and littoral seas.
- Create rear-area sanctuaries for U.S. forces through air superiority.
- Closely monitor enemy activities while denying the enemy the ability to do the same.
- Begin combat operations in the manner and at the time and place the United States chooses.
- Seize the initiative with a massive air and missile campaign focused on achieving air superiority in the opening hours or days.
- Maintain the offensive initiative through parallel and continuous air operations throughout the depth of the battlespace.
- Sustain the air campaign from air bases and carriers generating large numbers of aircraft sorties (intelligence, surveillance, and reconnaissance [ISR], strike, refueling) with industrial efficiency and uninterrupted by enemy action.

This way of war served the United States well in the past two decades. As is the case with any period of military dominance, however, new technologies and concepts eventually reduce imbalances between opposing force capabilities. This appears to be happening today because of the proliferation of long-range strike and other systems:

- **Emerging long-range strike capabilities are bringing the era of sanctuary to an end, with significant implications for the American way of war.** Highly accurate long-range strike systems, particularly ballistic and cruise missiles, make it much harder to have useful rear-area sanctuaries because these missiles are extremely difficult to defend against and can reach those locations from which U.S. forces would prefer to operate. This preference is driven by simple geography. The United States relies primarily on relatively short-ranged fighter aircraft for most tactical air missions (e.g., close air support, interdiction, air superiority, suppression of enemy air defenses), and, although fighters can sustain operations from as far as 1,500 nautical miles (nm), bases within 500 miles of operating areas are greatly preferred. Similarly, ground forces have a limited ability to transit long distances over land and generally depend on major ports within 500 miles of their operating areas. Advances in man-portable precision strike systems, although a lesser threat, also raise the possibility that adversary special operations forces (SOF) could be more lethal in the future than in past conflicts. For these reasons, against some adversaries, U.S. forces will deploy to bases that are at risk of attack from the onset of an operation. A period of buildup under conditions of sanctuary can no longer be taken for granted against the most capable adversaries.

- **The United States will need to adapt its power-projection concepts to operate under a greater threat of attack.** Power-projection concepts will likely change in several ways. Without sanctuary, U.S. forces might need to deploy in a more deliberate and cautious manner, treating rear areas as combat zones and moving in a tactical, rather than administrative, fashion. Defensive forces and preparations will, by necessity, become a higher priority, likely delaying the movement of some offensive capabilities into theater. Deployments might be much less efficient as forces are moved in smaller waves through multiple, dispersed ports and airfields. An enemy attack during the deployment phase could further slow down and disrupt force movements and might make it difficult for the United States to gain the initiative, at least in the early phases of combat. Prepositioning of logistics support equipment and changes to current support processes could improve the USAF's ability to deploy to dispersed locations and operate under attack.

- **As in the past, a combination of measures is needed, but the specific mix will vary depending on the political geography of the region, adversary capabilities, and U.S. objectives.** Active defenses, hardening, CCD, dispersal, and postattack recovery all make unique contributions to air base defense and operability. If resources were unlimited, defenders would want to fully avail themselves of all these options at every operating location. In the real world, however, choices will have to be made. Extensive hardening, for example, might be a good choice at main operating bases where multiyear peacetime infrastructure projects are feasible. In contrast, hardening options are going to be much more limited at expeditionary locations because of time constraints, if nothing else. In either case, hardening is not a panacea because it cannot be effective against all threats, and the more-permanent structures make it possible for adversaries to do detailed peacetime targeting against these known locations. Similarly, although CCD possibilities should be considered at all locations, the best fit might be at dispersal or auxiliary bases where hardening is more problematic but CCD can exploit enemy preconceptions about how the United States will use such bases (e.g., how often and under what conditions forces will move among them). Operations research techniques can identify the optimal mixes of each capability at individual bases and across the theater. That said, in a major conflict, adversary actions will disrupt optimal resource allocations, and, in many cases, local base defenders will have to improvise. A good grasp of best practices and adaptations from previous wars will be valuable in those situations.

Recommendations

This research leads to three primary recommendations for USAF and U.S. Department of Defense planners:

- **Consider the air base, the airspace above and near it, and the surrounding land as a battlespace, a place where defenders cannot expect sanctuary.** Too often, base defense and recovery are treated as support functions to be delegated to security forces and civil engineers. Although base and wing commanders take base defense seriously, it has not been a priority for the institutional air force, primarily because it has not been conceptualized as a core warfighting problem. It also has not received the attention and resources from the joint community, a critical problem because ground-based air defense of air bases is an Army responsibility. The relatively low priority for air base defense has led to a variety of shortfalls in USAF capabilities and in Army ground-based air defense capabilities. Understanding the air base as a battlespace makes its defense a core mission for the USAF and should help build a consensus among senior leaders to push forward new concepts, doctrine, and capabilities for operating under attack, possibly including the creation of USAF air and missile defense units.

- **Develop and test new concepts of operation for deployment to and operation of air bases under attack that incorporate both historical lessons and a full appreciation of emerging threats.** The American way of war as developed and executed between 1990 and 2011 is beginning to fray. Although current practices might remain viable for many years against weaker foes, it is time to begin developing alternative concepts and capabilities for more-contested environments in which disruptions to lines of communication, resource constraints, and local improvisation might be the rule rather than the exception. The U.S. experience during 1942 and early 1943 is particularly instructive because U.S. forces routinely operated under (and overcame) such conditions. New concepts can be developed and tested at relatively low cost using existing forces and infrastructure as the USAF did in 2014 with its Rapid Raptor concept.[3] For example, tests could explore the practicality of dispersing aircraft at varying intervals across existing ramp space at a single base, dispersing small packages of aircraft across multiple bases, or periodically moving units among a larger set of bases.

- **Explore organizational options to better support distributed and dispersed operations.** Although the mix of air base defense measures will vary by theater and threat, dispersed operations are likely to become central to operational concepts against highly capable adversaries. The full exploitation of dispersed concepts might, therefore, require changes to USAF organization and support structures. Today, these are designed to maximize economies of scale by operating large forces from relatively few main operating bases. An organization built around the air wing and intended to operate the wing at a single location (or at most two locations) might not be well suited for combat environments that require small force elements (e.g., squadron size or smaller) to be widely dispersed among many locations.

---

[3] Michael Trent Harrington, "Rapid Raptor Moves JBER F-22s Closer to the Fight," Joint Base Elmendorf-Richardson, May 15, 2014; referenced September 4, 2014.

Acknowledgments

I thank the following RAND colleagues who have been particularly influential (over many years) in advancing my understanding of the anti-access problem (both political and operational) and air base attack and defense specifically: Roger Cliff, Natalie Crawford, Paul Dreyer, David Frelinger, Jeff Hagen, Ted Harshberger, Eric Heginbotham, Jacob Heim, Andrew Hoehn, Sherrill Lingel, Michael Lostumbo, Forrest Morgan, David Ochmanek, Chad Ohlandt, David Orletsky, Stacie Pettyjohn, David Shlapak, William Stanley, Donald Stevens, Paula Thornhill, and Barry Wilson. Thanks also to Col Shannon Caudill, Lt Col Russell Davis, and Col Michael Pietrucha, U.S. Air Force, for their detailed and helpful suggestions on an earlier draft. Reviewers John Drew and David Ochmanek of RAND and Thomas Mahnken of the Naval War College provided expert, constructive, and helpful reviews. Finally, thanks to Lisa Bernard for her skilled and careful edit of the manuscript.

Abbreviations

A2	anti-access
AAA	anti-aircraft artillery
AB	air base
APS	Army prepositioned stock
CCD	camouflage, concealment, and deception
CEP	circular error probable
DoD	U.S. Department of Defense
DPICM	dual-purpose improved conventional munition
DPRK	Democratic People's Republic of Korea
GPS	Global Positioning System
IAF	Indian Air Force
IRBM	intermediate-range ballistic missile
ISR	intelligence, surveillance, and reconnaissance
JBB	Joint Base Balad
MANPADS	man-portable air defense system
MOB	main operating base
MRBM	medium-range ballistic missile
NATO	North Atlantic Treaty Organization
nm	nautical mile
NVA	North Vietnamese Army
OAF	Operation Allied Force
ODF	Operation Deliberate Force
ODS	Operation Desert Storm
OEF	Operation Enduring Freedom
OIF	Operation Iraqi Freedom
OTH	over the horizon
PACAF	Pacific Air Forces
PAF	RAND Project AIR FORCE
PLA	People's Liberation Army
RAF	Royal Air Force
RNAS	Royal Navy Air Service
SOF	special operations forces
SRBM	short-range ballistic missile
TAB VEE	Theater Air Base Vulnerability Evaluation Exercise
TBM	tactical ballistic missile

TEL	transporter-erector-launcher
TLAM-D	Tomahawk Land Attack Missile–Dispenser
USAAF	U.S. Army Air Forces
USAF	U.S. Air Force
USAFE	U.S. Air Forces in Europe
USMC	U.S. Marine Corps
USN	U.S. Navy
VC	Vietcong

Chapter One. Introduction

Background

The U.S. Air Force (USAF) counts on access to overseas bases to project power abroad. However, today, it has only a fraction of the large main operating bases (MOBs) it maintained during the Cold War.[4] Peacetime MOBs remain a key component of security cooperation with the United States' closest partners and are vital enablers of global power projection, but, with a few exceptions, their direct warfighting role might be diminishing as threats to forward bases grow.

U.S. defense planners have to increasingly consider the possibility that operational access will be disrupted through enemy military action. Although threats to operational access have been a factor in past conflicts,[5] the potential for an integrated anti-access system is relatively new.

Andrew Krepinevich, Barry Watts, and Robert Work first articulated the concept of anti-access and area denial in 2003 to describe a strategy that future opponents might use to defeat U.S. power-projection capabilities.[6] Krepinevich and his coauthors define anti-access strategies as those that "aim to prevent U.S. forces entry into a theater of operations," while area-denial operations "aim to prevent their freedom of action in the more narrow confines of the area under an enemy's direct control."[7] The anti-access dimension would include efforts to disrupt U.S. movement of forces to overseas theaters (e.g., interdiction of sea lines of communication), as well as attacks on forward ports, airfields, ground-force concentrations, and supporting infrastructure (which admittedly do more than just deny entry into the theater but, equally

---

[4] For a history of the U.S. military presence abroad, see Stacie L. Pettyjohn, *U.S. Global Defense Posture: 1783–2011*, Santa Monica, Calif.: RAND Corporation, MG-1244-AF, 2012.

[5] For example, German submarines impeded U.S. shipping access to England during much of World War II. Axis anti-access (A2) efforts, however, failed to hinder the October 1942 Operation Torch armada transiting from Hampton Roads, Virginia, to North Africa. Similarly, the Japanese South Pacific campaign sought to cut off U.S. access to Australia. The Japanese advance was turned back at the Battle of the Coral Sea. A month later, the Japanese experienced their first defeat at the Battle of Midway. Combined with the stalwart Allied defense of New Guinea and the U.S. landing and eventual victory at Guadalcanal, reliable U.S. access to the South Pacific was ensured for the remainder of the war.

[6] See Andrew F. Krepinevich, Barry D. Watts, and Robert O. Work, *Meeting the Anti-Access and Area Denial Challenge*, Washington, D.C.: Center for Strategic and Budgetary Assessments, 2003. These ideas were updated and expanded in two later Center for Strategic and Budgetary Assessments reports: Andrew F. Krepinevich, *Why AirSea Battle?* Washington, D.C.: Center for Strategic and Budgetary Assessments, 2010; and Jan van Tol, Mark Alan Gunzinger, Andrew F. Krepinevich, and Jim Thomas, *AirSea Battle: A Point-of-Departure Operational Concept*, Washington, D.C.: Center for Strategic and Budgetary Assessments, 2010. For a historical and conceptual treatment of A2 warfare, see Sam J. Tangredi, *Anti-Access Warfare: Countering A2/AD Strategies*, Annapolis, Md.: Naval Institute Press, 2013.

[7] Krepinevich, Watts, and Work, 2003, p. ii.

important, would also hinder operations). The area-denial dimension includes capabilities (e.g., advanced air defenses, naval mines, and antiship cruise missiles) intended to prevent U.S. forces from conducting air operations near or over the adversary's territory or maritime operations in its littoral seas.

The Policy Problem

Simply put, the United States is dependent on access to overseas bases to achieve political and military objectives. Future technologies (e.g., long-range strike, hypersonics, and reusable space planes) might lessen U.S. military dependence on overseas bases, but no concept or capability in research and development is anywhere near mature enough (or cheap enough) to enable warfighting exclusively from intercontinental distances in the near future. Thus, for the next decade or more, the United States will be dependent on military access to overseas bases. Yet, changes in the international political and military environment are likely to make such access more difficult to gain and sustain in both peacetime and during conflict.[8] Thus, the key challenge facing policymakers and planners is to develop a global access strategy that achieves political access where needed for peacetime and contingency requirements and that ensures that U.S. forces can achieve operational objectives despite enemy attacks on air bases, ports, and other forward facilities.

Purpose of This Report

This report is intended as a reference on air base attack and defense to inform public debate, as well as government deliberations, on what has become known as the A2 problem, specifically as it applies to air base operations.

Organization of This Report

Chapter Two presents a brief review of the history of airfield attacks. Chapter Three considers how U.S. attention to this threat waned after the Cold War ended and a new American way of war formed in this lower-threat environment. Chapter Four describes emerging missile and ground threats to air base operations and considers whether these threats represent disruptive innovations. Chapter Five discusses defensive counters to air base attack, both as used historically and as future options. Chapter Six presents conclusions and recommendations.

---

[8] For an assessment of challenges to and options for U.S. global basing, see Stacie L. Pettyjohn and Alan J. Vick, *The Posture Triangle: A New Framework for U.S. Air Force Global Presence*, Santa Monica, Calif.: RAND Corporation, RR-402-AF, 2013.

Chapter Two. A Short History of Attacks on Air Bases

From October 8, 1914, when the Royal Navy Air Service (RNAS) first successfully attacked the German Zeppelin base at Dusseldorf, to the August 2014 Islamic State of Iraq's capture of Tabqa Air Base (AB) in Syria,[9] air forces have lost aircraft parked on the ground to attack by air, sea, and ground forces. In the intervening 100 years, air bases have been attacked by land- and sea-based fighter and bomber aircraft; by ballistic and cruise missiles;[10] by naval gunfire;[11] by artillery, mortars, and rockets; and by small teams of commandos, insurgents, or terrorists who penetrated perimeter defenses.[12] Air bases have also been overrun during major ground offensives, although documentation of aircraft losses is largely missing for such battles. In contrast, there is good documentation of aircraft losses from smaller ground attacks. These alone damaged or destroyed more than 2,000 aircraft between 1941 and 2014.[13]

Although impressive, these losses pale in comparison to the damage that U.S. air forces did to Axis airfields during World War II. U.S. Army Air Forces (USAAF), U.S. Navy (USN), and U.S. Marine Corps (USMC) aircraft reportedly destroyed more than 18,000 enemy aircraft on the

---

[9] On this date, a single RNAS Sopwith Tabloid aircraft dropped two 20-lb. Hale bombs on one of two Zeppelin sheds near Dusseldorf, destroying one Zeppelin. See Ian Castle, *The Zeppelin Base Raids: Germany 1914*, Oxford, UK: Osprey Publishing, 2011, pp. 22–26. Ironically, this attack was immediately followed by one of the first ground attacks on an airfield, with German artillery hitting the Antwerp airfield where the RNAS Sopwith had launched (Castle, 2011, p. 27). The first documented (although unsuccessful) attack on an airfield was six weeks earlier on August 24, 1914, when a Royal Flying Corps patrol aircraft dropped a bomb on a German airfield near the town of Lessines, Belgium. See John F. Kreis, *Air Warfare and Air Base Air Defense, 1914–1973*, Washington, D.C.: Office of Air Force History, 1988, pp. 5–6; and Ruth Sherlock, "Isil Fighters Capture Key Syrian Air Base in Sign of Growing Strength," *Telegraph*, August 24, 2014; referenced September 1, 2014.

[10] The United States has used cruise missiles in airfield attacks in at least one operation. During Operation Allied Force, cruise missiles struck the commercial and military airfield at Batajnica, Serbia. See Benjamin S. Lambeth, *NATO's Air War for Kosovo: A Strategic and Operational Assessment*, Santa Monica, Calif.: RAND Corporation, MR-1365-AF, 2001, p. 21. The Navy developed the Tomahawk Land Attack Missile–Dispenser (TLAM-D) cruise missile (carrying 166 Combined Effects Munitions) during the Cold War specifically to attack Soviet aircraft parked in the open. See Ronald O'Rourke, *Cruise Missile Inventories and NATO Attacks on Yugoslavia: Background Information*, Washington, D.C.: Congressional Research Service, April 20, 1999.

[11] Japanese battleship and cruiser bombardment of Henderson Field was a nightly occurrence during the most desperate days of fighting on Guadalcanal. See Eric M. Bergerud, *Touched with Fire: The Land War in the South Pacific*, New York: Penguin Books, 1996, pp. 391–394, 463; and Thomas Guy Miller, *The Cactus Air Force*, New York: Harper and Row, 1969.

[12] An excellent historical treatment of the evolution of air base air defense is Kreis, 1988. For a comprehensive study of air warfare in the South Pacific during World War II, see Eric M. Bergerud, *Fire in the Sky: The Air War in the South Pacific*, Boulder, Colo.: Westview, 2001. Bergerud devotes considerable space to the exploration of both the tactics and strategic significance of air base defense and attack in the Pacific campaign. For a history of ground-force attacks on air bases, see Alan J. Vick, *Snakes in the Eagle's Nest: A History of Ground Attacks on Air Bases*, Santa Monica, Calif.: RAND Corporation, MR-553-AF, 1995.

[13] Most of these losses occurred between 1940 and 1975. See Vick, 1995, p. 19.

ground during World War II alone.[14] We could not find summary statistics for Allied losses on the ground to Axis air attack, but there is good documentation showing that roughly 1,500 Allied aircraft were lost to Axis attacks during the Battle of Britain, from Japanese strikes against airfields on Oahu, and during Operations Barbarossa and Bodenplatte (see Table 2.1). Additionally, in 1940, the French Air Force might have lost a few thousand aircraft destroyed or captured on the ground, and the low countries likely lost dozens.[15] The one country that lost relatively few combat aircraft (although many trainer aircraft) on the ground was Poland. Despite the popular image to the contrary, the Poles dispersed their fighter forces to makeshift runways the day before the German attack and suffered few losses on the ground.[16] Given the scale and length of the war and frequency of air attacks on Allied airfields in all theaters, several thousand additional Allied aircraft might have been destroyed on the ground. Total combatant aircraft losses on the ground during World War II could easily have reached 25,000. Table 2.1 lists details for some of the campaigns and conflicts that saw the heaviest attacks on airfields in the past 100 years.

---

[14] The USAAF reported destroying 12,040 enemy aircraft on the ground during World War II. See U.S. Air Force, *United States Air Force Statistical Digest World War II*, Washington, D.C.: Management Information Division, Directorate of Management Analysis, Comptroller of the Air Force, Headquarters, U.S. Air Force, December 1945; referenced September 24, 2013, pp. 263–268. According to the USN, the number of enemy aircraft destroyed on the ground between December 1941 and August 1945 (in all theaters) by USN and USMC aircraft is 6,182. See Office of Naval Intelligence, Air Branch, *Naval Aviation Combat Statistics: World War II*, Washington, D.C.: Naval Aviation History Office, June 17, 1946, p. 18.

[15] Norway, Denmark, and the Netherlands had very small air forces; a few dozen aircraft were probably captured or destroyed on the ground during the German invasion in 1940. A much larger number of French aircraft was destroyed on the ground or captured during the Battle of France in 1940, perhaps a few thousand. The French Air Force possessed 4,360 aircraft as of May 1940 but "could only bring into action one-fourth of the aircraft available." This suggests that many were lost on the ground. See Faris R. Kirkland, "The French Air Force in 1940: Was It Defeated by the Luftwaffe or by Politics?" *Air University Review*, September–October 1985; referenced September 29, 2014.

[16] Steve Zaloga, *Poland 1939: The Birth of Blitzkrieg*, Oxford, UK: Osprey, 2002, p. 50.

Table 2.1. Aircraft Destroyed or Damaged on the Ground, Selected Campaigns and Conflicts, 1940–1999

Date	Conflict or Operation	Airfield Attacker	Airfield Defender	Type of Attack	Aircraft Lost
August–September 1940	Battle of Britain	Germany	Great Britain	Air	56
June 1941	Operation Barbarossa	Germany	Soviet Union	Air	800
December 1941	Attack on Pearl Harbor	Japan	United States	Air	347
1941–1943	North African Campaign	British special forces	Germany	Commando	367
January 1945	Operation Bodenplatte	Germany	United States, United Kingdom, Canada	Air	388
1942–1945	World War II (all theaters)	USN, USMC, USAAF	Axis countries	Air	18,222
1965	India–Pakistan War	Pakistan and India	India and Pakistan	Air	IAF: 35, Pakistan Air Force: 9
June 1967	Six-Day War	Israel	Egypt, Syria, Jordan, Iraq	Air	400
1964–1973	Vietnam War	NVA, VC	United States	Mortar and rocket	1,578
1965–1972	Vietnam War	USAF	North Vietnam	Air	163
January–February 1991	ODS	Allied coalition	Iraq	Air	151
March–May 1999	OAF	NATO	Serbia	Air	100

SOURCES: Aircraft-loss data are from the following sources. For Battle of Britain, James, 2000, Appendix 34. For Operation Barbarossa and the attack on Pearl Harbor, Higham and Harris, 2006, pp. 272, 288. For North Africa, Vick, 1995, p. 19. For Operation Bodenplatte, Manrho and Pütz, 2010, p. 461. For World War II, U.S. Air Force, 1945, and Office of Naval Intelligence, 1946, p. 18. For the 1965 India–Pakistan War: Jagan Mohan and Chopra, 2006, Appendixes B and C. For the Six-Day War, Kreis, 1988, p. 319. For U.S. losses in Vietnam, Vick, 1995, p. 19. For North Vietnamese losses, Thompson, 2000, Table 5, p. 307. For Operation Desert Storm, DoD, 1992, p. 206. For Operation Allied Force, DoD, 2000, p. 69.

NOTE: IAF = Indian Air Force. NVA = North Vietnamese Army. VC = Vietcong. ODS = Operation Desert Storm. OAF = Operation Allied Force. NATO = North Atlantic Treaty Organization.

This review of conflicts between 1914 and 2014 found that airfields were attacked by ground forces, air forces, or both at least 26 times, as shown in Table 2.2. What is striking about the examples displayed in the table is that combatants consider airfields important targets not just in large wars between major powers but even in small conflicts, such as the four-day-long Soccer War between Honduras and El Salvador in 1969, and insurgencies, such as the ongoing conflict in Syria. This historical review is not intended to suggest that future attacks will use the same weapons and tactics but rather to establish how common and significant they were in past conflicts.

Table 2.2. Airfield Attacks, 1914–2014

Date of Attack	Conflict or Operation	Airfield Attacker	Type of Attacks on Airfield
1914–1918	World War I	United Kingdom, France, Germany, United States	Air
1936–1939	Spanish Civil War	Republicans, nationalists	Air
1939–1945	World War II	United Kingdom, United States, Russia, Germany, Japan	Air and ground
1950–1953	Korean War	DPRK, United States	Air and ground
1956	Suez crisis	United Kingdom, France	Air
1964–1973	Vietnam	United States, NVA, VC	Air and ground
1965	India–Pakistan War	India, Pakistan	Air
1967	Six-Day War	Israel	Air
1969	Soccer War	Honduras, El Salvador	Air
1971	India–Pakistan War	India	Air
1973	Yom Kippur War	Egypt, Israel	Air
1980–1988	Iran–Iraq War	Iraq, Iran	Air
1982	Falkland conflict	RAF, British Special Forces	Air and ground
1985	Angolan civil war	South Africa	Air
1986	Soviet Afghan intervention	Afghan insurgents	Ground
1987	Libya–Chad border war	Chad	Ground
1982, 1990	El Salvador civil war	Frente Farabundo Martí para la Liberación Nacional, or Farabundo Martí National Liberation Front, insurgents	Ground
1991	ODS	United States, United Kingdom, France, Kurdish insurgents	Air and ground
1992	Philippine insurgency	New People's Army	Ground
1994	Operation Deny Flight	NATO	Air
1999	OAF	NATO	Air
2001	Sri Lankan civil war	Liberation Tigers of Tamil Eelam	Ground
2001–2013	OEF	United States, Taliban	Air and ground
2003–2011	OIF	United States, Iraqi insurgents	Air and ground
2011	Operation Odyssey Dawn	NATO	Air
2011–2014	Syrian civil war	Syrian rebels	Ground

SOURCE: Authors' analysis of historical sources (see list of references at the end of this report).
NOTE: DPRK = Democratic People's Republic of Korea. RAF = Royal Air Force. OEF = Operation Enduring Freedom. OIF = Operation Iraqi Freedom.

For the United States, post–World War II combat experiences began to undermine the expectation that U.S. airfields would be vulnerable to attack in every conflict. During the Korean War, airfield attack was highly lopsided. North Korean Yakovlev (Yak-9) fighters managed a few attacks on USAF airfields on the first day of the conflict (June 25, 1950), damaging seven Republic of Korea trainer aircraft at Seoul airfield and destroying a USAF C-54 transport at Kimpo airfield. But when North Korean IL-10 fighters attempted airfield attacks again a couple

6

of days later on June 27, USAF F-82 fighters easily shot them down or drove them off. Although the North Korean Air Force continued to conduct offensive operations until July 20, it could not inflict any significant damage to friendly airfields in the first month of the war.[17] The North Korean Air Force did manage to conduct harassing attacks against Kimpo airfield in October 1951 but was largely quiet for the next year. Beginning in October 1952 and continuing sporadically for several months, DPRK PO-2 biplane trainers conducted harassing attacks against a variety of targets in the Seoul area, including airfields. The low- and slow-flying, fabric-covered aircraft proved difficult to detect and shoot down. In June 1953, these attacks were occurring most nights and became known as Bedcheck Charlie. United Nations (UN) forces went to considerable lengths to detect and defeat these attacks, but the attacks were never a serious threat to UN air operations.[18]

In contrast, the USAF conducted heavy attacks on North Korean airfields throughout the conflict, hindering DPRK air strikes on the south. That said, USAF attacks on DPRK airfields had no effect on communist jet fighter operations. These were conducted from bases in Manchuria, where they enjoyed political sanctuary and were never attacked by UN forces, which feared escalating the conflict.[19]

During the Vietnam War, the USAF attacked North Vietnamese airfields multiple times between 1965 and 1968. In response, the North Vietnamese flew most of their fighters to sanctuary in China. The March 1968 bombing halt, however, led the North Vietnamese Air Force to return the aircraft to Vietnam, and their airfields were once again fully operational by October 1968.[20] The North Vietnamese never attacked U.S. airfields from the air (likely because of a combination of limited capacity and the recognition that they had little to gain and much to lose by attacking U.S. airfields). This reinforced the view that, if air superiority could be achieved, friendly airfields would largely be free from air attack. On the other hand, North Vietnamese and VC ground forces attacked U.S. and Vietnamese air bases almost 500 times during that conflict. Primarily using mortars and rockets for standoff attacks, they destroyed 393 aircraft and damaged another 1,185 over a ten-year period.[21] These attacks were sufficiently worrisome that the USAF created a program, dubbed Concrete Sky, to design and construct hardened aircraft shelters for Vietnam.[22]

---

[17] Robert Frank Futrell, *The United States Air Force in Korea: 1950–1953*, Washington, D.C.: Office of Air Force History, 1983, pp. 7, 12–13, 101.

[18] Futrell, 1983, pp. 309–310, 661–665.

[19] See Kreis, 1988, p. 271.

[20] Kreis, 1988, pp. 279–296. See also Benjamin S. Lambeth, *The Transformation of American Air Power*, Ithaca, N.Y.: Cornell University Press, 2000.

[21] Vick, 1995, p. 19.

[22] These shelters are discussed in more detail in Chapter Five. See Roger Fox, *Air Base Defense in the Republic of Vietnam, 1961–1973*, Washington, D.C.: Office of Air Force History, 1979, pp. 71–73; and Karen Weitze, *Eglin Air*

While the United States was occupied in Vietnam, four wars in the Middle East and South Asia saw significant attacks on airfields. The experience from the two Arab–Israeli wars (1967 and 1973) and the two India–Pakistan wars (1965 and 1971) was mixed. On the one hand, airfields were attacked in every case and with spectacular success in 1967. On the other hand, airfield attacks were largely one sided in the Arab–Israeli wars and limited to the opening hours and days during all four conflicts. In 1965, the IAF lost 35 aircraft on the ground while the Pakistan Air Force lost nine. In 1971, the IAF had success in attacks on Pakistani runways, but neither side lost any military aircraft on the ground.[23] During the 1967 Arab–Israeli War, Israel decimated Arab air forces, destroying 400 (mainly Egyptian, but also Jordanian, Syrian, and Iraqi) aircraft in brilliant preemptive strikes. In 1973, Israel suffered some attacks on its airfields, but both conflicts appeared to validate the view that, once a combatant gained air superiority, that combatant's airfields were safe from attack.

Both NATO and Warsaw Pact nations expected that, in the event of a war in Europe, airfields would be heavily attacked.[24] The Warsaw Pact faced threats from NATO aircraft delivering conventional or nuclear weapons and, toward the end of the Cold War, nuclear-armed Pershing II intermediate-range ballistic missiles (IRBMs) and ground-launched cruise missiles. NATO faced a more complex threat, including Warsaw Pact aircraft, tactical ballistic missiles (TBMs) armed with conventional, chemical, or nuclear weapons and special forces.

During the 1950s, there was great concern about the vulnerability of NATO air bases to nuclear attack. This led to the development of the dispersed operating concept and various other programs designed to limit damage from both nuclear and conventional attacks.[25] Air base defense efforts included dispersing aircraft across a large number of bases and the construction of hardened aircraft shelters (beginning in the late 1960s), fuel storage, and command posts; active defenses; and airfield damage–repair capabilities. These initiatives were given a final

---

Force Base, 1931–1991: Installation Buildup for Research, Test, Evaluation and Training, Eglin Air Force Base, Fla.: Air Force Materiel Command, 2001, pp. 239–240.

[23] P. V. S. Jagan Mohan and Samir Chopra, *The India–Pakistan Air War of 1965*, New Delhi: Manohar, 2006, Appendixes B and C; P. V. S. Jagan Mohan and Samir Chopra, *Eagles over Bangladesh: The Indian Air Force in the 1971 Liberation War*, Noida, India: HarperCollins Publishers India, 2013, pp. 145, 151, 188–192.

[24] Air base attack and defense received sustained analytical attention during the Cold War, including the development of computer simulations, such as RAND's Theater Simulation of Airbase Resources and air base damage assessment models. See, for example, C. Richard Neu, *Attacking Hardened Air Bases (AHAB): A Decision Analysis Tool for the Tactical Commander*, Santa Monica, Calif.: RAND Corporation, R-1422-PR, 1974; Donald E. Emerson, *An Introduction to the TSAR Simulation Program: Model Features and Logic*, Santa Monica, Calif.: RAND Corporation, R-2584-AF, 1982; and Donald E. Emerson, *AIDA: An Airbase Damage Assessment Model*, Santa Monica, Calif.: RAND Corporation, R-1872-PR, 1976. For broader discussions of defensive options, see John Halliday, *Tactical Dispersal of Fighter Aircraft: Risk, Uncertainty, and Policy Recommendations*, Santa Monica, Calif.: RAND Corporation, N-2443-AF, 1987; and Bruce W. Don, Donald E. Lewis, Robert M. Paulson, and Willis H. Ware, *Survivability Issues and USAFE Policy*, Santa Monica, Calif.: RAND Corporation, N-2579-AF, 1988.

[25] Lawrence R. Benson, *USAF Aircraft Basing in Europe, North Africa, and the Middle East, 1945–1980*, Ramstein Air Base, Germany: Headquarters, U.S. Air Forces in Europe, 1981; declassified 2011 by Air Force History Office.

boost of energy in the mid-1980s with the Salty Demo air base survivability program, which developed new techniques for air base operations under attack. The ground threat to NATO bases (particularly from Soviet Spetsnaz) also received renewed attention in the 1980s. In their 1984 "31 Initiatives" memorandum of agreement, Army Chief of Staff GEN John A. Wickham and USAF Chief of Staff Gen Charles A. Gabriel specified several areas in which the Army would assist in the defense of USAF bases.[26] When the Cold War ended in 1989, the most severe threats to USAF bases evaporated.

After the lopsided U.S. victory over Iraq in 1991, threats to U.S. air bases seemed to many a thing of the past. The complete U.S. dominance of the air suggested a new era in which U.S. airpower could operate from forward bases without the risk of enemy attack. The only hint of possible future trouble came in the form of the Iraqi mobile conventional ballistic-missile force. This relatively small force was equipped with inaccurate Scud or Iraqi-modified Scuds that had no serious military value. Iraqi Scuds nevertheless terrorized Saudi cities, threatened to bring Israel into the war, and diverted a large number of USAF sorties in unsuccessful attempts to find and destroy transporter-erector-launchers (TELs). In a stroke of incredibly bad luck for U.S. forces, one Iraqi Scud hit a U.S. Army barracks facility, killing 28 soldiers—more fatalities than on any other day of that conflict.[27] There also was a close call at the port of Al Jubayl, where an incoming Scud broke up over the harbor and hit the water near "a large pier where six ships and two smaller craft were tied up" and 500 feet from an ammunition storage area on the pier.[28]

That said, the one Iraqi missile that did hit an air base (Dhahran) produced nothing more than a crater in an open area and might have been deflected there by a Patriot interceptor.[29] If anything, the Scud impact at Dhahran seemed to confirm the more sanguine view that conventional ballistic missiles were too few and too crude to present a serious threat to USAF operations. Absent a world-class adversary air force, it seemed that USAF bases would enjoy sanctuary and need not bother with the expensive and elaborate air base defense measures of past wars.[30]

---

[26] For more details, see Shannon W. Caudill, *Defending Air Bases in an Age of Insurgency*, Maxwell Air Force Base, Ala.: Air University Press, 2014, p. 363; and Richard G. Davis, *The 31 Initiatives: A Study in Army–Air Force Cooperation*, Washington, D.C.: Office of Air Force History, U.S. Air Force, 1987.

[27] Richard Hallion, *Storm over Iraq: Air Power and the Gulf War*, Washington, D.C.: Smithsonian Institution Press, 1992, pp. 185–187.

[28] Bernard Rostker, *Information Paper: Iraq's SCUD Ballistic Missiles*, Washington, D.C.: U.S. Department of Defense, interim paper, July 25, 2000.

[29] On January 20, 1991, a warhead or large part of a Scud missile impacted at Dhahran AB, creating a large crater in an open area. According to DoD, a Patriot missile intercepted the Scud. This interception presumably would have disturbed the Scud's original trajectory. This, combined with the very low accuracy of the Iraqi Scuds, makes it difficult to determine the intended target of the attack. See Rostker, 2000.

[30] As is discussed later in the report, not everyone in the defense community embraced this sanguine view of the threat. For example, the 1993 *Bottom Up Review* recognized the potential threat of cruise and ballistic missiles. See Les Aspin, *Report on the Bottom-Up Review*, Washington, D.C.: U.S. Department of Defense, October 1993.

The next chapter argues that, during ODS, the United States discovered a new way of war that became the template for all major power-projection operations during the 1990s and 2000s.

Chapter Three. A New American Way of War?

ODS ushered in a new "American Way of War"[31] in which land- and sea-based airpower, using stealth, precision, and large numbers of cruise missiles, rapidly dismantled a large and sophisticated Iraqi integrated air defense system; destroyed the Iraqi Air Force in the air and on the ground; severely limited Iraqi commanders' understanding of the battlespace; then, having gained air dominance, systematically attacked Iraqi armor dug in on the Kuwait–Saudi border. Whether or not airpower was the decisive element in the defeat of the Iraqi Army,[32] it certainly was decisive in creating the benign conditions that allowed a massive U.S. joint force to deploy, create elaborate logistical support infrastructure, and move two Army corps hundreds of miles to forward staging areas, all unmolested by Iraqi offensive action.[33] The air campaign quickly put an end to any Iraqi prospects for effective offensive operations, as demonstrated by the Iraqi offensive debacle at the Battle of Khafji, launched 12 days after the start of the air campaign.[34] The fact that U.S. and allied airpower made adversary offensive operations nearly impossible and created a level of sanctuary neither expected in 1990 nor experienced in previous conflicts was lost in the subsequent decisiveness-of-airpower debate. This is unfortunate because this was both a revolutionary development and the essential foundation for a new way of war that would dominate U.S. military operations for the next two decades.

---

[31] In his classic *The American Way of War: A History of United States Military Strategy and Policy* (Bloomington, Ind.: Indiana University Press, 1977), Russell Frank Weigley originated the idea of a unique American way of war. Weigley used it to capture the essence of U.S. grand strategy over the course of the nation's history. Numerous books have followed with similar titles, and their authors have used the phrase "American way of war" to capture other distinctive aspects of U.S. strategy or military operations, such as the U.S. emphasis on technology or the use of firepower to reduce U.S. casualties. See, for example, Thomas G. Mahnken, *Technology and the American Way of War Since 1945*, New York: Columbia University Press, 2008. Finally, Benjamin S. Lambeth concludes that the transformation of U.S. airpower between the Vietnam War and Operation Allied Force created a new American way of war in which the relative roles of air and ground power against large mechanized forces had changed, with airpower now the dominant killing mechanism. See Lambeth, 2000, especially pp. 313–320.

[32] Among the more-prominent works arguing that airpower largely crippled Iraqi ground forces prior to the allied ground campaign are Lambeth, 2000; Hallion, 1992; and David A. Deptula, *Effects-Based Operations: Change in the Nature of Warfare*, Arlington, Va.: Aerospace Education Foundation, 2001. This also was the judgment of the Gulf War Air Power Survey. For an alternative view arguing that the low quality of Iraqi ground forces was the decisive factor, see Daryl G. Press, "The Myth of Air Power in the Persian Gulf War and the Future of Warfare," *International Security*, Vol. 26, No. 2, Fall 2001, pp. 5–44. A related view offered by Stephen Biddle is that the interaction between Iraqi ineptitude and new U.S. technologies best explains the one-sided outcome. See Stephen Biddle, "Victory Misunderstood: What the Gulf War Tells Us About the Future of Conflict," *International Security*, Vol. 21, No. 2, Fall 1996, pp. 139–179.

[33] Iraq did take offensive action in the form of Scud missile attacks on Israel and Saudi Arabia, but these are best understood in political, rather than military, terms. They were designed to accomplish two purposes: (1) Provoke an Israeli retaliation that might make it impossible for Arab nations to remain in the coalition, and (2) undermine support for the war in Saudi Arabia. They presented no real military threat to allied forces.

[34] Lambeth, 2000, pp. 121–124.

This new American way of war[35] was replicated in Operations Deliberate Force (ODF), OAF, OEF and OIF in their opening phases, and, to a lesser degree, Operation Odyssey Dawn.[36] Although the specifics varied (particularly the role of ground forces), these operations all shared the following seven components:

- Rapidly deploy large joint forces to forward bases and littoral seas.
- Create rear-area sanctuaries for U.S. forces through air superiority.
- Closely monitor enemy activities while denying the enemy the ability to do the same.
- Begin combat operations in the manner and at the time and place the United States chooses.
- Seize the initiative with a massive air and missile campaign focused on achieving air superiority in the opening hours or days.
- Maintain the offensive initiative through parallel and continuous air operations throughout the depth of the battlespace.
- Sustain the air campaign from sortie factories—air bases and carriers generating large numbers of aircraft sorties (intelligence, surveillance, and reconnaissance [ISR]; strike; refueling) with industrial efficiency and uninterrupted by enemy action.[37]

This is not to suggest that these components capture every dimension of U.S. military operations since 1990; they clearly do not. These operations are, however, distinct from Cold War expectations and earlier combat experience because they share these unique characteristics, all of which are founded on the near invulnerability of friendly rear areas (including air bases and aircraft carriers) to enemy attack. The American-way-of-war argument takes no position on the relative effectiveness of air, land, and sea power. It does, however, recognize that recent operations have been heavily dependent on airpower, particularly in the opening phases. Finally, this line of analysis makes no judgments about whether the operations ultimately achieved U.S. strategic objectives. The primary objective in developing the idea of a new American way of war is to highlight its creation and dependence on rear-area sanctuary and, consequently, its fragility

---

[35] This section presents ideas first developed in a 2008 study for the USAF and widely briefed between 2008 and 2014. See Alan J. Vick, *Challenges to the American Way of War*, presentation to the Global Warfare Symposium, Los Angeles, Calif., November 17, 2011.

[36] For ODF, see Robert C. Owen, ed., *Deliberate Force: A Case Study in Effective Air Campaigning—Final Report of the Air University Balkans Air Campaign Study*, Maxwell Air Force Base, Ala.: Air University Press, 2000; and Lambeth, 2000, pp. 174–178. For Operation Allied Force, see Tony Mason, "Operation Allied Force: 1999," in John Andreas Olsen, ed., *A History of Air Warfare*, Washington, D.C.: Potomac Books, 2010, pp. 235–262; and Lambeth, 2001. For Operation Enduring Freedom, see Benjamin S. Lambeth, *Air Power Against Terror: America's Conduct of Operation Enduring Freedom*, Santa Monica, Calif.: RAND Corporation, MG-166-1-CENTAF, 2006. For OIF, see Benjamin S. Lambeth, *The Unseen War: Allied Air Power and the Takedown of Saddam Hussein*, Annapolis, Md.: Naval Institute Press, 2013. For Operation Odyssey Dawn, see DoD, "DoD News Briefing with Vice Adm. Gortney from the Pentagon on Libya Operation Odyssey Dawn," news transcript, March 19, 2011; referenced September 3, 2014.

[37] Although air base operability studies have understood and analyzed the air base as an industrial process for decades, John Stillion might have coined the term *sortie factory* in 2008 while on staff at RAND. He discussed it in a later interview with *Air Force Magazine*. See Marc V. Schanz, "Rethinking Air Dominance," *Air Force Magazine*, July 2013, p. 38.

in the face of adversaries that possess the motivation and capacity to execute effective attacks against vital rear areas. The following pages expand and develop each of the seven components of the new American way of war.

Rapidly Deploy Large Joint Forces

The ability to deploy large forces relatively quickly to critical regions over transoceanic distances is unique to the United States and central to its role as a global military power. The U.S. military executes this mission with such skill and apparent ease that Americans and many others largely take it for granted. This is quite remarkable given the scale and daunting complexity of peacetime, crisis, and wartime mobility operations and is a tribute to the American genius for military logistics. Indeed, no other nation in the world possesses strategic lift that even remotely compares to U.S. global mobility capabilities. These global capabilities include a network of overseas bases and access and overflight agreements (which provide airfields, air routes, ports, prepositioned equipment, and communication links), strategic airlift, air refueling aircraft, and strategic sealift.[38]

The modern American way of war counts on global mobility and begins to move forces forward during a crisis to signal resolve, enhance deterrence, and expand near-term military options for national leaders. These forces include USAF fighters and bombers, carrier strike groups, other naval assets (e.g., nuclear attack submarines operating independently), Marine expeditionary units, and light air-transportable Army units. If large ground forces are required, the next arriving deployments could be Army or USMC personnel deploying by air to link up with either land- or ship-based prepositioned equipment.[39] For major ground operations, such as ODS, the bulk of ground forces must deploy by ship from the continental United States, a process that can take months.

Create Rear-Area Sanctuaries

A notable aspect of this modern American approach to power projection is the creation of rear-area sanctuaries. These sanctuaries make three vital contributions. First, sanctuary enables

---

[38] Strategic sealift includes roll-on/roll-off, bulk cargo, container, fuels, and Army prepositioned stock (APS) ships.

[39] Prepositioning has been a critical component of the U.S. global military posture since early Cold War days, when the concept of prepositioning of materiel configured to unit sets was developed in Europe to speed the deployment of reinforcements from the United States. Today, APSs are located abroad at Camp Darby, Italy; Camp Arifjan and Kuwait Naval Base, Kuwait; Camp As Sayliyah, Qatar; Camp Carroll, South Korea; and Yokohama Harbor and Sagami Army Depot, Japan. There is also an APS afloat at Diego Garcia. See Association of the United States Army, "Army Prepositioned Stocks: Indispensable to America's Global Force-Projection Capability," Arlington, Va., December 2008; referenced April 3, 2014. There are also many prepositioning ships for the other services, known as maritime prepositioning ships at multiple locations. These contain munitions and a wide range of equipment and supplies to support USMC, USAF, USN, and U.S. Army operations. See U.S. Military Sealift Command, "Strategic Sealift (PM3)," undated; referenced April 4, 2014.

rapid and efficient deployment to and through airfields and ports. If the enemy were able to effectively strike these transportation nodes, deployments could be disrupted, diverted, and slowed by airfield or port closures while attacks were under way. Damage to airfield or port infrastructure (e.g., runways, fuel storage, piers, or cargo handling) would also hinder operations, and repair activities would likely exacerbate such effects, at least in the short run. Damaged or destroyed ships or aircraft would further complicate repair and recovery efforts while reducing the available lift capacity through attrition. The second benefit from sanctuary is the ability to base a large force at each airfield (as many as several hundred aircraft), making the most efficient use of the infrastructure and support assets. Doing this requires parking large numbers of aircraft relatively close together in the open, something that one would do only under conditions of sanctuary. The final benefit that sanctuary brings is the ability to launch missions shortly after the first forces arrive. This is critical for land-based air operations because they would often be the first forces employed in ISR, defensive, or offensive operations. That said, sanctuary is a powerful enabler for ground operations as well, providing some breathing space for recently arriving units to reintegrate personnel, vehicles, and equipment at marshaling areas and then to transition to tactical movements and subsequent defensive positions or offensive maneuver. Once naval forces begin conducting combat operations, they too are dependent on forward ports for rearming and refitting.

Given the great and immediate value of sanctuary, it should be no surprise that defensive forces typically have been among the first to deploy. If air and missile defenses were not already provided by U.S. forces permanently based in the area of operation, then the earliest deployments would include air and missile defense forces (e.g., USAF Airborne Warning and Control System and fighter squadrons, Army Terminal High Altitude Area Defense and Patriot battalions, and USN Aegis-equipped destroyers or cruisers).

Closely Monitor Enemy Activities

As noted above, the rapid deployment of advanced air defenses has allowed the United States to create air superiority up to (and within) adversary borders. Combined with the creation of rear-area sanctuary, this enables U.S. airborne ISR platforms to conduct operations over key operational areas and, in recent operations, up to adversary borders. In some cases (e.g., OIF and ODF), U.S. ISR platforms were operating over adversary terrain even prior to the specific named operation.[40] In contrast, U.S. air superiority has been so extensive in recent operations that adversary air forces have been unable to conduct any airborne ISR missions. Admittedly, the Afghan Air Force (such as it was) had no capacity to do this, but the Iraqi and Serbian air forces possessed some (admittedly modest) ISR capacity that they could not employ.

---

[40] This was highly contingent on the type of aircraft and sophistication of enemy air defenses (e.g., manned nonstealthy ISR platforms, such as the Joint Surveillance Target Attack Radar System, did not overfly Kosovo or Serbia prior to or during Operation Allied Force).

Begin Combat Operations

A frequently overlooked advantage that the United States exploited in conflicts between 1991 and 2011 was the freedom to choose the time, place, and manner in which U.S. forces entered combat. Although the United States was typically reacting to some kind of aggression by the adversary state, the conflicts did not start with enemy offensive action against U.S. forces. Rather, the United States was able to build up forces in theater (taking many months, in several cases) and then initiate combat when and where it was most favorable for U.S. and allied forces. This was a tremendous advantage and far from the conditions expected to prevail in a Cold War conflict with the Soviet Union in Central Europe or in a second Korean War. In both those cases, U.S. forces most likely would have suffered heavy losses from the opening enemy offensives and been forced to fight, perhaps for days or weeks, from a defensive and reactive position.[41] As U.S. planners look to the future, there are potential conflicts (e.g., a Chinese invasion of Taiwan) in which U.S. forces would also likely begin combat on the defensive rather than the offensive.

Seize the Initiative with Air and Missile Campaigns

As noted earlier, ODS was the model for this new way of war and began with the largest U.S. air operation since World War II. Early in the morning of January 17, 1991, air elements from all services launched a massive and highly coordinated attack focused primarily on dismantling the Iraqi air defense system. More than 2,700 sorties were flown in the first 24 hours, supporting close to 1,200 strikes against Iraqi air defense–sector headquarters, radars, air bases, military headquarters in Baghdad, national communications, and their fielded army.[42] Benefiting greatly from advances in stealth, precision, and electronic warfare, the allied air offensive dismantled the Iraqi command and control network and destroyed key elements of the air defense system while blinding or confusing the rest. Those Iraqi fighter aircraft that did make it into the air that first night were overmatched by allied crews, aircraft, and the USAF integrated airborne battle management system. Of the 25 Iraqi fighters that launched on January 17, eight were shot down.[43] U.S. Central Command concluded that air superiority had been achieved by the end of the first 24 hours, and GEN H. Norman Schwarzkopf, Jr., the U.S. Central Command commander, declared ten days later on January 27 that the coalition had achieved air supremacy,

---

[41] For a compelling fictional treatment that captures the expected intensity of a war in Europe, see John Hackett, *The Third World War: August 1985*, New York: MacMillan Publishing Company, 1979. Similarly, for a war in Korea, David Shlapak wrote a short but powerful vignette describing a fictional North Korean special forces attack on a U.S. air base in South Korea. See David A. Shlapak and Alan J. Vick, *"Check Six Begins on the Ground": Responding to the Evolving Ground Threat to U.S. Air Force Bases*, Santa Monica, Calif.: RAND Corporation, MR-606-AF, 1995, pp. 1–6.

[42] Thomas A. Keaney and Eliot A. Cohen, *Gulf War Air Power Survey: Summary Report*, Washington, D.C.: U.S. Air Force, 1993, pp. 12–13.

[43] Benjamin S. Lambeth, *The Winning of Air Supremacy in Operation Desert Storm*, Santa Monica, Calif.: RAND Corporation, P-7837, 1993, p. 4.

"meaning that the Iraqi Air Force no longer existed as a combat-effective force."[44] OIF also started with an air and missile campaign but not to achieve air superiority.[45] The first night, 1,700 sorties were flown, supported by an additional 504 cruise-missile strikes.[46] ODF, OAF, OEF, and Operation Odyssey Dawn all followed this model as well, albeit on a much smaller scale because of the different political objectives and operational conditions that they faced.[47]

Sustain the Air Campaign from Sortie Factories

All recent major U.S. military operations, including prolonged counterinsurgency operations in Iraq and Afghanistan, have depended on the ability of joint air elements to generate a large number of air missions—ISR, strike, close air support, air refueling, airlift, and aeromedical evacuation—day and night without interruption. The lethality, precision, persistence, and availability of close air support has enabled ground forces to move more quickly and operate in smaller and lighter force elements than they otherwise would. And the combination of tactical medevac out of the field with USAF aeromedical evacuation (including airborne life support, nurses, and other critical care) to world-class trauma facilities in Germany has saved countless lives.

Thus, a key element of the modern American way of war is the expectation that air missions will fly without interruption by enemy action. Indeed, one has to go back to World War II to find examples of enemy actions substantially disrupting U.S. air operations. The limited air attacks during the Korean War and even much larger number of ground attacks during the Vietnam War did little to disrupt U.S. sortie generation.[48] Similarly, Iraqi insurgent and Taliban mortar and rocket attacks on air bases, although quite common, did minimal damage to aircraft and failed to disrupt air operations. Even the most successful Taliban attack of the war—the September 2012 commando raid on Camp Bastion (which destroyed six Marine Corps Harriers and damaged another two)—had no impact on sortie generation or the ability of U.S. air elements to provide support to forces in the field.

---

[44] Keaney and Cohen, 1993, pp. 56–57.

[45] Air superiority had already been achieved through ODS and the no-fly zone and 1998–2002 suppression-of-enemy-air-defenses campaign that preceded OIF. Thanks to Michael W. Pietrucha for pointing this out.

[46] Lambeth, 2013, p. 82.

[47] For details of these operations, see Lambeth, 2000; Lambeth, 2006; and DoD, 2011.

[48] There are several reasons that the NVA and VC ground attacks on air bases did little to disrupt sorties. First, most attacks were quite small, with 93 percent of standoff attacks against USAF MOBs firing fewer than 40 rounds. Second, the NVA and VC were unable to sustain attacks over multiple days. For example, at Da Nang, the most frequently attacked MOB, only 16 of 95 attacks occurred within 48 hours of a previous attack. The interval between attacks allowed bases to recover. Finally, attacks were relatively infrequent, occurring on less than 2 percent of the days that MOBs operated over the course of the war. There were roughly 36,500 days when USAF MOBs might operate (ten MOBs times approximately ten years of conflict times 365 days = 36,500). Air base attacks occurred on just under 500 of these, leaving 36,000 unmolested base operating days. See Vick, 1995, Chapter Five, especially pp. 100–101.

This way of war was enabled by capabilities that the United States developed during the Cold War but, as noted previously, had little in common with how a major conflict with the Soviet Union was expected to unfold. The new American way of war served the United States well in major power-projection efforts between 1990 and 2011, but its viability in the future can no longer be taken for granted. In the next chapter, we explore how emerging capabilities in some potential adversary nations are beginning to undermine this power-projection concept.

Chapter Four. Back to the Future: The End of the Sanctuary Era

Between 1990 and 2014, the United States went to war against regional or lesser powers that had no means to challenge U.S. air superiority and therefore no means to threaten the American way of war. That attractive state of affairs possessed a superficial permanence similar to earlier dominant military concepts and technologies, such as the Greek phalanx, longbow, and battleship—all eventually overtaken by tactical or technological advances. Similarly, today, it appears that the proliferation of long-range precision strike capabilities, accurate and reliable ballistic and cruise missiles especially, is bringing an end to the American era of rear-area sanctuary.[49]

Ballistic and Cruise Missiles

In principle, air bases have been at risk of ballistic- or cruise-missile attack since the German development of the V-2 and V-1 missiles in the 1940s. In practice, these early missiles lacked the accuracy to reliably hit anything smaller than a large urban area and, thus, were used by the Nazi regime primarily for domestic political purposes—that is, to shore up civilian morale by attacking British cities (and, later, liberated cities in Belgium, the Netherlands, and France) in retaliation for Allied bombing of German population centers.

Using captured German V-2 technology, the Soviet Union developed and, in 1955, deployed the R-11 (known in the west by the NATO code name SS-1b Scud-A). This missile was the first in a family of TBMs that would be deployed in large numbers and exported to Warsaw Pact nations and other Soviet client states, such as Iraq and Libya. The R-11 carried a nuclear, chemical, or unitary conventional high explosive warhead (950 kg). The low accuracy of this version (3,000 m circular error probable [CEP]) did, however, limit its effectiveness as a conventional airfield attack weapon. The R-17 (Scud B) version entered service in 1962, offering much better accuracy (450 m CEP) and a greatly expanded menu of munition options, including unitary high explosive, submunition (including runway busting, fragmentation, armor-piercing, and mines), and chemical warheads.[50] Although NATO planners expected airfields to be hit by missiles in a war, they viewed the chemical munition–armed Scud as more problematic than the conventional version, which, even with its improved accuracy, was not a threat to hardened

---

[49] It comes as no surprise to students of military innovation that the endless cycle of measure–countermeasure in military affairs would eventually end the era of sanctuary. For an insightful overview of the military innovation literature, see Adam Grissom, "The Future of Military Innovation Studies," *Journal of Strategic Studies*, Vol. 29, No. 5, 2006, pp. 905–934.

[50] "R-11/-17 (SS-1 'Scud'/8A61/8K11/8K14, and R-11FM [SS-N-1B])," *Jane's Strategic Weapons Systems*, updated January 7, 2014; referenced April 21, 2014.

facilities; Warsaw Pact fighters, such as the MiG-23 Flogger, were expected to do the most physical damage to NATO airfields.[51]

NATO had viable defensive counters to Cold War–era threat systems and embraced them all. Larger aircraft (e.g., tankers) that were impractical to protect in hardened shelters were based in Spain and England, outside the range of Soviet fighter aircraft and Scud missiles. Hardened aircraft shelters, underground fuel storage, and command-post bunkers were constructed at NATO fighter bases, providing good protection from small submunitions and all but a direct hit from a large unitary warhead (unlikely, given the accuracy of Warsaw Pact munitions). NATO air defenses included fighter aircraft, long- and short-range surface-to-air missiles. These were counted on to severely attrite attacking Warsaw Pact aircraft. Finally, air base recovery capabilities were greatly expanded so that runways could be rapidly repaired and operations resumed.[52]

The advent of Global Positioning System (GPS)–guided weapons in the 1990s raised the possibility that, in future conflicts, ballistic missiles might present a much greater threat to airfields than was expected during the Cold War or experienced during ODS. The speed of ballistic missiles offers an attacker two inherent advantages: (1) short flight time (which minimizes warning) and (2) a high probability of penetrating defenses, particularly when launched in large waves. Once ballistic missiles achieved acceptable accuracy, they had the potential to become the weapon of choice for airfield attack, particularly at the beginning of a short war and especially for attackers inclined toward preemptive strategies.

John Stillion and David Orletsky at RAND conducted the first detailed analysis of the effects of more-accurate ballistic- or cruise-missile attacks on airfields. Their 1999 report assessed the vulnerability that aircraft parked in the open had to attacks by TBMs armed with unitary warheads or submunitions.[53] They focused on the threat to aircraft in the open because, from ODS on, the U.S. military often parked large numbers of aircraft relatively close together on open ramps without even revetments for protection.[54] Their analysis used USAF spacing guidelines to estimate parking densities as inputs for a Monte Carlo simulation.[55]

Figure 4.1 is reproduced from the Stillion and Orletsky report and shows several striking features. First, note, in the lower left corner, the small circle showing the lethal area for an M-9

---

[51] Christopher J. Bowie, "The Lessons of Salty Demo," *Air Force Magazine*, March 2009, pp. 54–57.

[52] Bowie, 2009. See also Stephen C. Hall, "Air Base Survivability in Europe: Can USAFE Survive and Fight?" *Air University Review*, September–October 1982; referenced June 8, 2014.

[53] John Stillion and David T. Orletsky, *Airbase Vulnerability to Conventional Cruise-Missile and Ballistic-Missile Attacks: Technology, Scenarios, and U.S. Air Force Responses*, Santa Monica, Calif.: RAND Corporation, MR-1028-AF, 1999.

[54] Shelters and revetments were used when available in those conflicts, but large aircraft were parked on open ramps, and even fighters were parked close together in the open at some bases.

[55] From Secretary of the Secretary of the Air Force, *Facility Requirements*, Washington, D.C., Air Force Handbook 32-1084, September 1, 1996, Table 2.6.

(now called the DF-15) missile carrying a single 500-kg unitary warhead, which would damage only six aircraft. In contrast, consider the largest circle in the figure.[56] This is the lethal radius for the M-9 (DF-15) ballistic missile if, instead of a single 500-kg warhead, its payload were used to carry 825 1-lb. submunitions. For relatively soft targets, such as aircraft on a large parking ramp, many small submunitions are vastly more efficient in creating lethal effects than a single big warhead. In this case, the bomblets damage 82 out of the 95 aircraft parked on the notional ramp (compared with six aircraft damaged by the single 500-kg warhead).

Figure 4.1. Comparison of Warhead Lethal Radii for Ballistic and Cruise Missiles

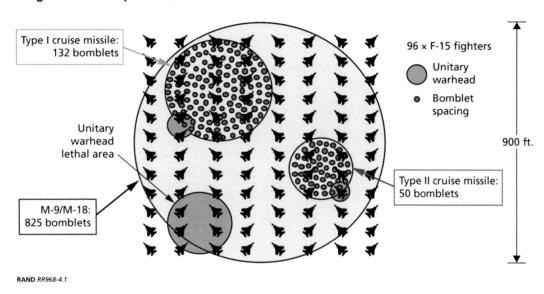

RAND RR968-4.1

SOURCE: Stillion and Orletsky, 1999, Figure 2.3, p. 14.

NOTE: Aircraft spacing reflects guidelines set out in Secretary of the Air Force, 1996, Table 2.6, for F-15 aircraft parked at a 45-degree angle.

We should note that the utility of submunitions against soft or area targets is not a recent discovery. The effectiveness of these weapons has been recognized at least since World War II, when both Allied and Axis powers used air-dropped cluster bombs, such as the U.S. M29.[57] As discussed above, the Soviet Union equipped its 1960-vintage Scud missiles with submunitions, and the United States developed a large family of antipersonnel and anti-armor submunitions for delivery by aircraft or artillery, cruise missile (TLAM-D), or battlefield rocket (multiple-launch rocket system or Army Tactical Missile System).[58] The United States has used cluster munitions

---

[56] Note that *M-18* refers to a notional missile that was never developed or fielded.

[57] See "M29 Cluster Bomb," National Museum of the U.S. Air Force, February 4, 2011; referenced May 26, 2014.

[58] Equipped with 166 BLU-97/B Combined Effects Bomblets (each weighing 1.5 kg), the TLAM-D was specifically designed for attacks against Soviet aircraft parked in the open. See "RGM/UGM-109 Tomahawk," *Jane's Strategic Weapon Systems*, updated May 6, 2014; referenced May 26, 2014. The U.S. Army multiple-launch rocket system, according to Jane's, carries "644 M77 Dual-Purpose Improved Conventional Munition (DPICM) shaped-charge

in every conflict since World War II, delivering millions over the course of the past 70 years.[59] The Chinese also possess a large and varied inventory of cluster munitions and, in their open literature, have highlighted the utility of these weapons for airfield attack.[60]

What was new with the Stillion and Orletsky report was analytical evidence that combining cluster munitions with accurate missiles greatly increased the lethality and efficiency of missile attacks on airfields. Stillion and Orletsky simulated attacks on four of the bases that the USAF used during ODS (Dhahran, Doha, Riyadh, and Al Kharj) and found that an attack with roughly 100 missiles could achieve a 0.9 probability of kill against aircraft parked in the open at these bases. To cover the vast parking areas (equivalent to almost 1,000 football fields, or 44 million square feet), they allocated 30 GPS-guided M-9 and 30 M-18 ballistic missiles (both Chinese designs) and 38 small GPS-guided cruise missiles.[61] This analysis demonstrated that, once ballistic-missile accuracy improved to the point that the CEP was around 150 m, such air base attacks would become feasible.[62] Missiles with this accuracy would also be effective in attacks on runways and taxiways. These attacks, using larger submunitions that penetrate the concrete then detonate underneath, create many craters, heaving of concrete, and considerable blast debris. Such attacks are designed both to trap aircraft on the base (allowing for their destruction by follow-on strikes) and to prevent sorties from being generated.[63]

There are now deployed systems with accuracies equivalent to or better than that postulated by Stillion and Orletsky. For example, Jane's reports that multiple variants of the DF-15 short-range ballistic missile (SRBM) have CEPs between 5 and 45 m, well below the Stillion and Orletsky threshold for effectiveness in airfield attacks.[64] Ballistic missiles do not yet have the accuracy to offer high-confidence destruction of hardened targets, such as aircraft shelters, but the proliferation of high-quality cruise missiles offers potential adversaries a highly

---

blast fragmentation bomblets weighing 213 g each." See "Lockheed Martin Missiles and Fire Control 227 mm Multiple Launch Rocket System (MLRS)," *Land Warfare Platforms: Artillery and Air Defence*, updated July 25, 2013; referenced May 26, 2014.

[59] See Geneva International Centre for Humanitarian Demining, *A Guide to Cluster Munitions*, Geneva, November 2007, p. 24. Other useful references on submunitions are Ove Dullum, *Cluster Weapons: Military Utility and Alternatives*, Oslo, Norway: Norwegian Defence Research Establishment, FFI-rapport 2007/02345, February 1, 2008; and Human Rights Watch, *Memorandum to CCW Delegates: A Global Overview of Explosive Submunitions*, prepared for the Convention on Conventional Weapons Group of Governmental Experts on the Explosive Remnants of War, May 21–24, 2002, Washington, D.C., 2002.

[60] Roger Cliff, John F. Fei, Jeff Hagen, Elizabeth Hague, Eric Heginbotham, and John Stillion, *Shaking the Heavens and Splitting the Earth: Chinese Air Force Employment Concepts in the 21st Century*, Santa Monica, Calif.: RAND Corporation, MG-915-AF, 2011, pp. 184–185.

[61] Stillion and Orletsky, 1999, p. xiv. The cruise missile was a notional design.

[62] Stillion and Orletsky, 1999, p. 9.

[63] For an analysis of the effects that runway-cutting attacks could have on the air war during a major conflict, see David A. Shlapak, David T. Orletsky, Toy I. Reid, Murray Scot Tanner, and Barry Wilson, *A Question of Balance: Political Context and Military Aspects of the China–Taiwan Dispute*, Santa Monica, Calif.: RAND Corporation, MG-888-SRF, 2009, pp. 37–44.

[64] "DF-15," *Jane's Strategic Weapon Systems*, updated June 5, 2014; referenced September 8, 2014.

complementary weapon system that possesses the accuracy and payload to achieve high probabilities of kill against aircraft shelters and other priority point targets. Cruise missiles, whether launched from the air, ground, or sea, can be used in precursor attacks or in parallel with or as follow-on attacks after ballistic-missile strikes, greatly complicating the air and missile defense problem.[65]

Effects of Missile Attacks on Air Bases

How consequential would such attacks likely be? One can imagine a wide range of outcomes, depending on the relative balance of capabilities between the adversary and U.S. forces. At one extreme, smaller, less accurate attacks conducted once or sporadically against hardened bases might result in modest aircraft or personnel losses and temporary damage to airfield infrastructure at one or a few bases. Such attacks might disrupt U.S. operations for a few hours at one or a few bases but would be unlikely to materially affect the prosecution of the war.

At the other extreme, larger and accurate attacks sustained over time against a less hardened posture could be devastating, causing large losses of aircraft and prolonged airfield closures. These effects were quantified in a 2009 RAND study of the China–Taiwan military balance. This analysis conducted combat simulations of Chinese missile attacks on Taiwan's airfields and found that,

> depending on missile accuracy, between 90 and 240 SRBMs—a number well within the range of estimates of the number of launchers China will field in the near future—could, with proper warheads, cut every runway at Taiwan's half-dozen main fighter bases and destroy essentially all of the aircraft parked on ramps in the open at those installations.[66]

Another RAND study found similar results for U.S. bases near China:

> [N]umbers on the order of 30–50 TBM per base appear to be sufficient to overload and kill air defenses, cover all of the open parking areas with submunitions to destroy aircraft parked there, and crater runways such that aircraft cannot [take off] or land.

> If we compare the numbers of missiles required to close bases with the numbers that China is currently fielding, clearly the U.S. could face extended periods of time when few, if any, of our bases near China are operating. RAND analysis has estimated that in the near-future, even with conservative estimates of TBM

---

[65] For a description of some of the more-advanced cruise-missile systems possessed by potential U.S. adversaries, see U.S. National Air and Space Intelligence Center, *Ballistic and Cruise Missile Threat*, Wright-Patterson Air Force Base, Ohio, 2013; referenced April 19, 2014.

[66] Shlapak, Orletsky, Reid, et al., 2009, p. xv.

production, Kadena could be kept closed to fighter operations for at least a week and kept closed to heavy aircraft . . . for much longer.[67]

In the midrange, there is the possibility of a more dynamic offense-defense competition, reminiscent of World War II, in which airfields are routinely attacked but active and passive defenses reduce the effectiveness of the attacks.[68] Under those conditions, airfields would continue to operate, albeit at a lower operating tempo, and the outcome of the battle of the airfields might be uncertain until the very end of the conflict. Although ballistic and cruise missiles present the most visible emerging threat to air bases, cheaper and simpler systems might also become more lethal in coming years. We now turn to one such threat: attack by adversary ground forces.

Ground-Force Threats to Air Bases

Since 2003, insurgent forces have launched at least 1,800 ground-force attacks on U.S. airfields in Iraq and Afghanistan, primarily with mortars or rockets. Standoff attacks have typically been quite small, with the attackers fleeing after firing a couple of rounds. For example, USAF data for attacks on Joint Base Balad in Iraq between 2004 and 2010 document that roughly 66 percent of attacks fired only one round and that approximately 87 percent of attacks fired one or two rounds. The largest attack recorded fired only 11 rounds.[69] Although attempts to directly assault or penetrate the perimeter of air bases were infrequent, the one spectacular success by insurgents was via commando assault. This was the September 14, 2012, commando attack on Camp Bastion, Afghanistan, that destroyed six USMC AV-8B Harrier jets and damaged another two. A dozen or so Taliban, dressed in U.S. Army uniforms, penetrated the perimeter and then used rocket-propelled grenades and hand grenades to destroy aircraft. All the Taliban attackers were killed in the attack.[70]

Although these attacks have caused loss of life and materiel, at no point have airfield attacks disrupted operations for extended periods, prevented air support for ground operations, or

---

[67] Jeff Hagen, *Potential Effects of Chinese Aerospace Capabilities on U.S. Air Force Operations*, testimony presented before the U.S.–China Economic and Security Review Commission on May 20, 2010, Santa Monica, Calif.: RAND Corporation, CT-347, 2010, pp. 2–3.

[68] The contested nature of what Norman Franks (1994) called "the battle of the airfields" is captured powerfully in Bergerud, 2001.

[69] The total number of attacks is not in the public record, but some reliable data are available from the 332nd Expeditionary Security Force Group for attacks on Joint Base Balad, Iraq, between 2004 and 2010. Standoff attacks for those years totaled roughly 1,800, with yearly numbers varying from approximately 130 in 2010 to approximately 475 in 2006. (Data are displayed in figures, not tables, so we had to estimate actual numbers.) See Joseph A. Milner, "The Defense of Joint Base Balad: An Analysis," in Shannon W. Caudill, ed., *Defending Air Bases in an Age of Insurgency*, Maxwell Air Force Base, Ala.: Air University, 2014, pp. 217–244, Figure 5.4, p. 232.

[70] Alissa J. Rubin, "Audacious Raid on NATO Base Shows Taliban's Reach," *New York Times*, September 16, 2012.

damaged large numbers of aircraft. Indeed, other than the Taliban commando attack on Camp Bastion in 2012, there were only three incidents in which mortar or rocket attacks destroyed or damaged aircraft in Iraq or Afghanistan, according to public record.[71] This is in contrast to the Vietnam War, during which there were roughly 500 ground attacks on U.S. and Vietnamese air bases over a comparable period (ten years). Only 21 of these attacks were commando-style, and they did relatively little damage. It was mortar and rocket attacks that destroyed nearly 400 aircraft and damaged close to 1,200.[72]

An example of the potential effectiveness of mortars when used against aircraft parked on a crowded ramp is found in the November 1, 1964, VC attack on Bien Hoa AB. During the night of October 31, elements of a VC company emplaced six 81-mm mortars approximately 1,400 m north of the B-57 parking ramp.[73] Over the course of a 20-minute attack that began shortly after midnight, the attackers fired somewhere between 50 and 65 rounds at multiple targets on the air base. They first struck the B-57 ramp, then a U.S. Army cantonment area, where four U.S. personnel were killed and 74 wounded.[74] As shown in Figure 4.2, 13 of the mortar rounds landed on the B-57 ramp, destroying five B-57s, causing heavy damage to eight aircraft and light damage to another seven.[75] An entire B-57 squadron was taken out of action in a single attack. The losses could have been much worse; two B-57 squadrons were parked on that small ramp until just nine days before the attack.[76]

---

[71] Four AH-64 Apaches were reportedly destroyed by mortar fire in Iraq in 2007, and one RAF Harrier was destroyed and one damaged by Taliban rocket attack in Afghanistan in 2005. See Cheryl Rodewig, "Geotagging Poses Security Risks," *www.army.mil*, March 7, 2012; and Sean Rayment, "Harrier Destroyed by Afghan Rocket," *Telegraph*, October 16, 2005. Although it caused minor damage, an August 2012 Taliban rocket attack on Bagram Airfield in Afghanistan was widely reported because it lightly damaged Chairman of the Joint Chiefs GEN Martin E. Dempsey's C-17 and slightly wounded two ground personnel. General Dempsey was not on the aircraft at the time. See Richard A. Oppel, Jr., and Graham Bowley, "Rocket Fire Damages Plane Used by Joint Chiefs Chairman," *New York Times*, August 21, 2012.

[72] Vick, 1995, pp. 19, 68.

[73] "Follow-Up to Bien Hoa Mortar Attack," Project CHECO staff report, Hickam Air Force Base, Hawaii: Headquarters Pacific Air Forces, December 1965; declassified by U.S. Air Force on January 9, 1991, p. 2.

[74] Impact points for 49 rounds were found and another possible ten to 15 impact points located. See Headquarters Military Assistance Command Vietnam, *Report of Investigation of Mortar Shelling of Bien Hoa Air Base on 1 November 1964*, Saigon, Vietnam, November 26, 1964; declassified by U.S. Air Force on January 9, 1991, p. 22.

[75] Richard R. Lee, *7AF Local Base Defense Operations, July 1965–December 1968*, Hickam Air Force Base, Hawaii: Headquarters Pacific Air Forces, July 1, 1969, p. 27.

[76] Maj Gen (later, Lt Gen) Joseph H. Moore, the 2nd Air Division commander, convinced 13th Air Force and PACAF headquarters to return one of the two squadrons to its home station at Clark AB, Philippines. Moore was particularly concerned about the vulnerability of the aircraft to mortar attack. In one exchange with the 13th Air Force commander, Moore cited a recent analysis that concluded that up to 50 percent of the aircraft could be lost to such an attack. See Kenneth Sams, *Historical Background to Viet Cong Mortar Attack on Bien Hoa: 1 November 1964*, Project CHECO (Contemporary Historical Examination of Current Operations) Office, Headquarters 2nd Air Division, November 9, 1964; declassified by U.S. Air Force on January 9, 1991, pp. 7–9. Also see Lee, 1969, p. 27. General Moore referred to two analyses of the mortar threat, one by 2nd Air Division and another by PACAF. The full package of PACAF, 5th Air Force, and 2nd Air Division staff memos and both analyses is found in Frederick

Figure 4.2. Mortar Impact Points, November 1, 1964, Attack on Bien Hoa Air Base, Vietnam

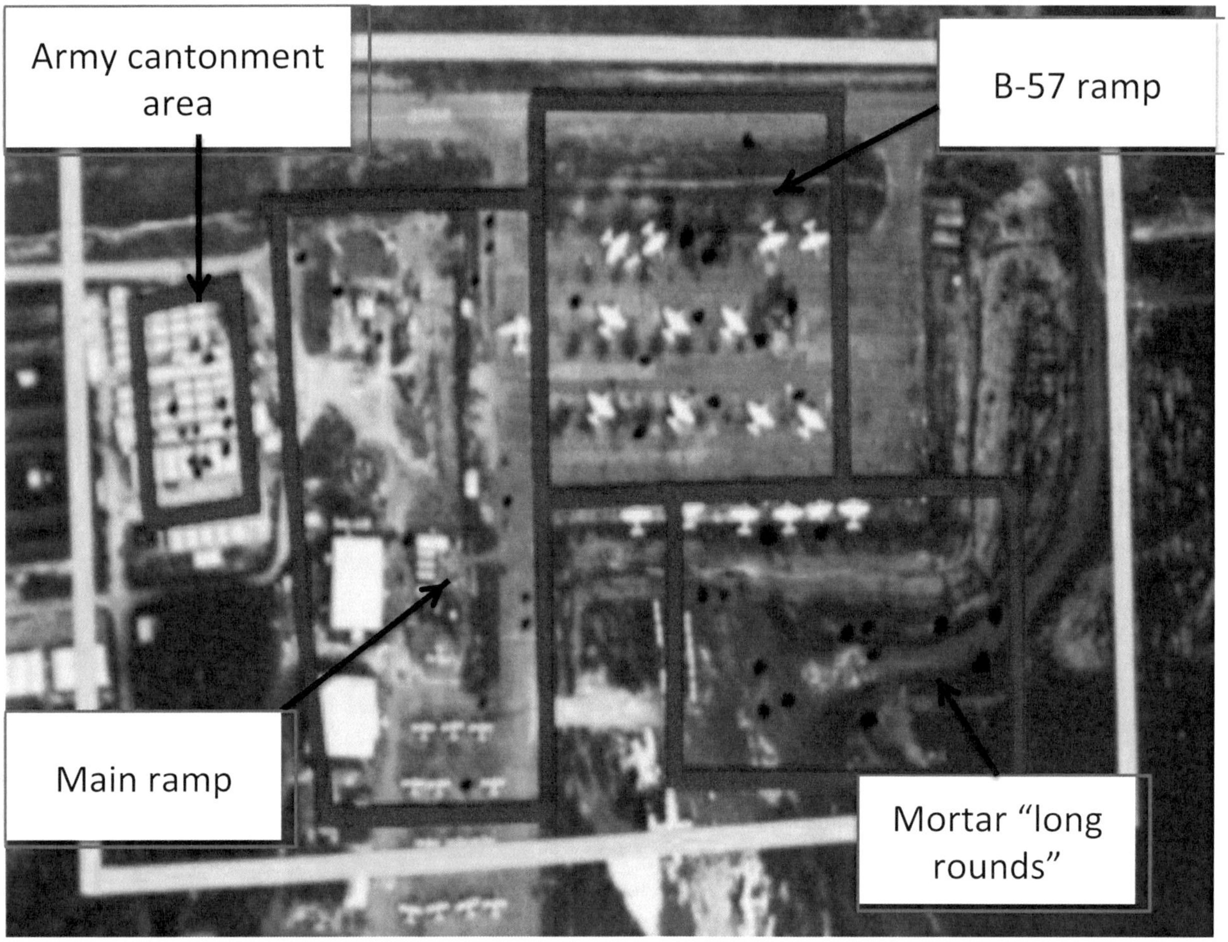

Army cantonment area

B-57 ramp

Main ramp

Mortar "long rounds"

Photo courtesy of Pacific Air Forces (PACAF) History Office.

NOTE: The photo was taken in the morning or afternoon of November 1, 1964. Informed by ground surveys, the impact points were noted with a black marker and four major impact areas noted with red markings. We added the text labels.

Unlike VC and North Vietnamese mortarmen, neither the Iraqi insurgents nor Taliban have managed a single spectacular mortar attack or a significant cumulative damage record against airfields. What accounts for the difference in effectiveness? Some argue that it was the overcrowding of USAF bases in Vietnam that made them ripe targets for attack. It is true that air bases in Vietnam were generally small and crowded, certainly compared with a vast base, such as Joint Base Balad (JBB), but ramps at Bagram Airfield in Afghanistan were sufficiently crowded that Vietnam War–style attacks with good mortar accuracies or massed rocket fires could have done considerable damage.

Torgerson, *Parked Aircraft Vulnerability to Mortar Attack*, Hickam Air Force Base, Hawaii: Headquarters Pacific Air Forces, September 1964; declassified by U.S. Air Force on August 23, 1990.

The answer appears to be twofold: Iraqi and Taliban attacks were generally quite small compared with those in Vietnam, and the skills of Iraqi and Taliban mortarmen varied greatly. For example, during the intense fighting in Ramadi in 2004, it was common for smaller bases, such as Camp Blue Diamond, to come under accurate mortar attack (typically four or five rounds) on a daily basis.[77] The average level of insurgent mortar crew skill was higher in Ramadi, the targets smaller, and U.S. ability to detect and counter standoff threats more limited. In contrast, MOBs, such as JBB, although attacked often, were generally in more-secure areas, were vastly larger, and appeared to have more resources devoted to defeating standoff attacks. For example, at JBB, the defensive strategy sought to prevent massed fires through a variety of deterrent and disruptive means. Also attackers were often low skilled or even novice subcontractors paid by insurgent forces and conducting attacks in a hurried way from unprepared positions.[78] U.S. forces took advantage of modern technologies and their superior ISR capabilities to prevent enemy forces from concentrating in sufficient force to do damage and, when attacks did occur, to rapidly defeat ongoing attacks.[79] Thus, unlike the VC, Iraqi insurgents were unable to adjust fire or sustain attacks long enough to do damage. As a result, these attacks rarely fired more than two mortar rounds.[80] Also, insurgents assigned to launch these small harassing attacks against Balad and other air bases appear to have been less skilled and often using older or even homemade mortars.[81] Finally, Iraqi and Taliban insurgents often used less accurate rockets, rather than mortars, in attacks against air bases.[82]

Future Ground Threats to Air Bases

Despite the relative ineffectiveness of Iraqi and Taliban forces in airfield attacks, there is growing concern that future adversaries might be more capable and effective. Three primary

---

[77] Thanks to RAND colleague Ben Connable for this observation and, more broadly, for sharing his experience and insights (and the following citation) regarding the effectiveness of Iraqi insurgent mortar attacks. For a report on a particularly lethal Iraqi mortar attack, see Matt Hilburn, "'Shock Wave': Seabees Recount Deadly Mortar Attack," *Navy Times*, May 31, 2004.

[78] Caudill, 2014, p. 203.

[79] Although we have no data for standoff attacks in Afghanistan, this description is consistent with what we know about attacks on Bagram AB and other MOBs in that country.

[80] Milner, 2014, Figure 5.4.

[81] Jim Michaels and Charles Crain, "Insurgents Showing No Sign of Letting Up," *USA Today*, August 22, 2004; referenced June 9, 2014. See also Martin Sieff, "Analysis: Mortar Attack Fits Deadly Pattern," United Press International, January 7, 2004; referenced June 9, 2014.

[82] In various conflicts, 57-mm, 107-mm, 140-mm, and 240-mm rockets have all been used in attacks on air bases. If used in large numbers, they can be quite effective and damaged or destroyed many USAF aircraft during the Vietnam War. They are, however, less accurate than mortars for at least two reasons. First, forward observers typically direct mortar fire, giving adjustments to a single mortar until the rounds are landing within lethal radius of the target. At that point, multiple mortars (using these refined settings) then fire en masse on the target. In contrast, small rockets suitable for infantry are single-use weapons, often fired from crude launchers, such as bamboo poles or pits.

threats deserve consideration: (1) unguided mortar attacks executed by well-trained mortar crews (e.g., major-nation special operations forces [SOF]);[83] (2) precision standoff attacks executed by adversary SOF, terrorists, or insurgents;[84] and (3) penetrating attacks by well-trained commandos.

Unguided Mortars

As demonstrated by the VC during the Vietnam War, skilled mortar crews using standard 81-mm mortars can strike with sufficient accuracy to destroy or severely damage many aircraft on crowded parking ramps. The VC successes were not flukes but a function of the inherent accuracy of mortars in the hands of skilled crews. In some cases, the VC were able to benefit from forward observers who adjusted fire onto targets; even without forward observers, though, these systems are dangerous. For instance, a typical military mortar can fire ten to 15 rounds for one minute without overheating. A relatively small attack using one to five mortars could, therefore, send ten to 75 rounds downrange in one minute. The attackers would essentially fire at maximum rate then head for the hills. Such an attack would take advantage of the inherent accuracy of the mortar and natural dispersal of rounds to cover an area target quickly, before base defenses could respond.[85] Well-trained conventional infantry or SOF are most capable of executing this type of attack, but there is no reason that future insurgents, terrorists, or mercenaries (particularly if they are former military) could not effectively use mortars in this way.

Another possibility is that unguided mortars could be made more lethal in unskilled hands by using mortar submunitions. As Stillion and Orletsky demonstrated in their analysis of submunition attacks on aircraft parking ramps, such warheads would greatly increase the lethality of mortar attacks. At least nine nations held mortar submunitions in their weapon inventories as recently as 2002.[86] An example is the Spanish MAT-120 cargo bomb, a 120-mm mortar round that carries 21 DPICM rounds. Although these submunitions were banned by the

---

[83] For a technical analysis of mortar and artillery threats to large aircraft on the ground, see Michael A. Silver, *Theater Airlifter Survivability on the Ground*, Wright-Patterson Air Force Base, Ohio: Air Force Institute of Technology, master's thesis, 1993.

[84] For an analysis of potential adversary tactics, see James Bonomo, Giacomo Bergamo, David R. Frelinger, John Gordon IV, and Brian A. Jackson, *Stealing the Sword: Limiting Terrorist Use of Advanced Conventional Weapons*, Santa Monica, Calif.: RAND Corporation, MG-510-DHS, 2007; and Shlapak and Vick, 1995.

[85] The inherent accuracy (without adjusted fire) of a U.S. Army 120-mm at maximum range (measured in CEP) is 108 m. This is sufficient for a few dozen mortar rounds to cause significant damage to aircraft parked in the open. The CEP of 108 m is derived from a formula indicating that the CEP for an unguided mortar round is approximately 1.5 percent of its range. At the mortar's maximum range of 7,200 m, the CEP is therefore 108 m. See Raymond Trohanowsky, U.S. Army Research, Development and Engineering Command Armament Research, Development and Engineering Center, "120mm Mortar System Accuracy Analysis," presented to International Infantry and Joint Services Small Arms Systems Annual Symposium, Exhibition, and Firing Demonstration, May 17, 2005.

[86] Belgium, China, France, Israel, Italy, Russia, Serbia, Spain, and Switzerland. For more on mortar submunition inventories, see Human Rights Watch, 2002.

2008 Convention on Cluster Munitions[87] and are no longer produced by Spain, they were exported to many countries, including Libya. During the Libyan civil war in 2011, three years after the cluster-munition ban, the Libyan Army used these weapons against rebel forces.[88] It is unclear how many nations still hold such weapons in their stockpiles or how much of the Libyan Army inventory has entered the global arms market, but it seems reasonable to assume that mortar submunition rounds are available to insurgents and others who want them.

Precision Standoff Attacks

Another possibility is that adversary or other ground forces would be equipped with advanced standoff systems and therefore be able to conduct precision strikes against parked aircraft and other high-value targets.[89] These could include guided rockets, artillery, missiles or mortars, small unmanned aerial vehicles, or large-caliber sniper rifles.[90] Guided mortar systems using GPS guidance are already deployed by the U.S. Army and in development by other nations.[91] These are likely to proliferate and, with CEPs under 20 m, have high hit probabilities against aircraft-size targets. Rafael Advanced Defense Systems' Spike NLOS (non–line-of-sight) missile system is an example of a small, ground-launched precision strike missile with a range of 25 km.[92] Although too expensive for many combatants and unlikely to be used in large numbers, Spike NLOS offers an extremely precise option to attack high-value aircraft in the open and could be used in the armed reconnaissance role, selecting its target while en route to the air base. Small armed unmanned aerial vehicles, such as AeroVironment's Switchblade, offer man-portable precision standoff for light infantry forces. Weighing only 5.5 lb., Switchblade has a 10-km range, carries a grenade-class munition, and has a beyond–line-of-sight video and control link. This link allows the operator to use Switchblade in a reconnaissance or strike role. For the latter, the operator simply flies Switchblade (remotely) into the target.[93]

---

[87] Diplomatic Conference for the Adoption of a Convention on Cluster Munitions, 108 signatories, August 1, 2010.

[88] The MAT-120 carries 21 DPICM rounds each weighing 275 g (50 g of explosive fill). Each round is designed to dispense submunitions across an area with a radius of 18 m. See "120 mm MAT-120 Cargo Bomb," *Jane's Infantry Weapons*, updated October 11, 2011; referenced September 12, 2013; and C. J. Chivers, "Qaddafi Troops Fire Cluster Bombs into Civilian Areas," *New York Times*, April 15, 2011.

[89] For a discussion of these technologies, see Bonomo et al., 2007, pp. 20–26. For a game-theory analysis of the precision standoff threat to airfields, see Jeffrey A. Vish, *Guided Standoff Weapons: A Threat to Expeditionary Air Power*, Monterey, Calif.: Naval Postgraduate School, master's thesis, 2006.

[90] A related threat that we do not analyze here is the man-portable air defense system (MANPADS) threat to aircraft in approach patterns or taking off. For additional details on the MANPADS threat, see Shlapak and Vick, 1995, pp. 52–54; and Caudill, 2014, p. 129.

[91] Audra Calloway, "Picatinny Fields First Precision-Guided Mortars to Troops in Afghanistan," *www.army.mil*, March 29, 2011; Richard Axford and Marcus Gartside, "Complex Weapons in a Time of Austerity," presented at Royal Aeronautical Society Conference, June 12, 2012; referenced September 13, 2013.

[92] See Rafael Advanced Defense Systems, "Spike NLOS™: Multi-Purpose, Multi-Platform Electro Optical Missile," undated; referenced May 26, 2014.

[93] AeroVironment, "Switchblade," V0I.1, 2012; referenced May 26, 2014.

Finally, large-caliber sniper rifles in the hands of trained marksmen are precision weapons capable of hitting targets the size of a human head at distances up to 1 mile and aircraft-size targets at distances up to 2 miles. Using a range of ammunition, including armor-piercing and incendiary, these rifles are a serious threat to aircraft parked in the open.[94] Direct-fire weapons, such as sniper rifles, do, however, require that the shooter have unobstructed line of sight to the parked aircraft, either from high terrain (e.g., hills or buildings) or at ground level (either on or off base). For these reasons, sniper rifles are less versatile than indirect-fire weapons but, where terrain or base layout permits, could present a significant threat to parked aircraft.[95]

Commando Attacks

Commando attacks represent another option to attack airfields. British Special Air Service commandos in North Africa used this tactic with great success during World War II, where they destroyed 367 German aircraft on the ground, primarily with satchel charges. The Special Air Service also used this tactic during the Falklands War in 1982, when it destroyed or damaged 11 Argentine aircraft on the ground at Pebble Island.[96]

SOF present the most severe threat of this type, given their training, mission-planning assets, and specialized transport, weapon, communication, and navigation systems. During the Cold War, NATO expected Soviet Spetsnaz to attack air bases in Germany using mortars, MANPADS, or penetrating techniques.[97] In some conflicts, such as a future Korean War, adversary SOF would likely play a substantial role in air base attacks because of their large numbers and the proximity of air bases.[98] The Chinese would also likely use SOF in any conflict against a technologically advanced foe. Writing in 2007, Roger Cliff and his coauthors observed,

> In addition to missile and air strikes, Chinese sources indicate that covert operatives, such as SOF or saboteurs, would also play an important role in attacks on enemy air bases. Some writings on the general missions that are assigned to SOF units suggest their role in air base attacks would include

---

[94] For more details on sniper-rifle threats to parked aircraft, see Shlapak and Vick, 1995, pp. 50–51; and John L. Plaster, *The Ultimate Sniper: An Advanced Training Manual for Military and Police Snipers*, Boulder, Colo.: Paladin Press, 1993, pp. 399–400. For details on the McMillan TAC-50, a representative large-caliber sniper rifle, see McMillan Firearms, *TAC-50 McMillan Tactical Rifle*, date unknown; referenced May 26, 2014. Bonomo et al., 2007, discusses ballistic computers and other advanced technologies that will increase the lethality of future sniper threats.

[95] Notwithstanding press reports to the contrary, sniper rifles are not practical weapons against aircraft in flight. For a discussion of this possibility, see Matthew L. Wald, "Citing Danger to Planes, Group Seeks Ban on a Sniper Rifle," *New York Times*, January 31, 2003; referenced June 9, 2014.

[96] Vick, 1995, pp. 17, 37–65.

[97] Director of Central Intelligence, *Warsaw Pact Nonnuclear Threat to NATO Airbases in Central Europe: National Intelligence Estimate*, Washington, D.C.: Central Intelligence Agency, NIE 11/20-6-84, October 25, 1984; declassified; referenced June 9, 2014. See also Erin E. Campbell, "The Soviet Spetsnaz Threat to NATO," *Airpower Journal*, Summer 1988; referenced June 9, 2014.

[98] See Shlapak and Vick, 1995, pp. 40–43, for a discussion of North Korean SOF capabilities.

strategic reconnaissance, harassment attacks, and direct-action missions, such as carrying out strikes on critical base facilities, destroying aircraft, and assassinating key personnel.[99]

SOF present a worrisome threat to individual air bases but also face significant constraints. One constraint is simple supply and demand: The potential targets for SOF attack could include ports, command posts, communication facilities, and airfields at minimum. The number of potential targets would likely exceed enemy SOF capacity in most major conflicts. Furthermore, in some scenarios, adversary SOF would have to travel great distances to attack U.S. targets and would face significant risks of detection and interception during transit and especially during their final approach to targets. Special operations are, by definition, high risk, and the attacker can have little certainty of its success, particularly for the most ambitious and complex missions. For these reasons, in most conflicts outside the Korean peninsula, SOF would likely play an important but more limited role, limited in both the number of targets they could attack and the frequency and size of these attacks. SOF would presumably be assigned to attack targets that enemy commanders deemed most important. For instance, they could be expected to attack critical assets at a few major bases, perhaps in conjunction with aircraft or missile strikes. For those facilities, additional defensive measures specifically focused on defeating the SOF threat are called for. These include high-quality friendly ground forces, improved perimeter sensors and barriers, additional hardening of key facilities, and dedicated airborne surveillance and fire-support platforms.

Commando-style attacks have also been conducted by a variety of combatants who often possessed fairly limited skills, suggesting that the threat is not exclusively from highly trained special forces. Puerto Rican separatists and Taliban, Tamil, Kurdish, and Philippine insurgents have all conducted successful commando-style attacks on airfields.[100] Insurgent and terrorist commando attacks on airfields could be relatively common in some conflicts, but the individual attack sophistication and potential for success would be much lower than for SOF. Good perimeter defenses and limited on-base rapid-reaction forces are generally adequate to defeat such threats, although the 2012 Camp Bastion attack is a reminder that, if there are lapses in defenses, even less capable adversaries can exploit them.

Effects of Ground Attacks on Air Bases

A single ground attack on an airfield, if well planned and executed, might destroy a large number of aircraft, cause significant personnel losses, and damage or destroy a few critical assets. Such an attack might be consequential, particularly in smaller conflicts, in which the political impact of a successful attack could turn public opinion against the operation. In a major

---

[99] Roger Cliff, Mark Burles, Michael S. Chase, Derek Eaton, and Kevin L. Pollpeter, *Entering the Dragon's Lair: Chinese Antiaccess Strategies and Their Implications for the United States*, Santa Monica, Calif.: RAND Corporation, MG-524-AF, 2007, p. 63.

[100] Vick, 1995, pp. 15–20; Rubin, 2012.

war, however, it is unlikely that the success of the U.S. war effort would turn on damage to a single base. In a major war, the U.S. adversary would have to destroy large numbers of aircraft and inflict substantial damage on runways and other airfield infrastructure at multiple locations to seriously disrupt U.S. operations. Moreover, attacks on air bases would have to be sustained over days and weeks across many airfields. This is difficult to do with missiles and aircraft, let alone light ground forces. The probability that small ground forces alone could create this level of damage and sustain it over time at many locations is low. This is not an excuse for complacency but does suggest that such attacks would likely play a supporting, rather than primary, role in adversary-nation strategies to counter U.S. airpower.

Disruptive Innovation and the American Way of War

Clayton M. Christensen and colleagues at Harvard Business School developed the concept of disruptive innovation to describe how market-dominant companies can lose market share or even be driven out of business by technological or business-model innovations that, at first blush, seem uncompetitive.[101] Using a series of compelling case studies, they identified the primary mechanisms of disruptive innovation. First, and somewhat surprising, the market-dominant company is often quite knowledgeable about the emerging technology. Indeed, in some cases, a company's own research and development department attempted to pursue the technology but was stymied by company management. This is because the emerging technology underperformed on the metrics that mattered most to the company and its clients. For example, Christensen describes the case of the 8-inch computer hard drive displacing the 14-inch hard drive, which, until the mid-1980s, was the standard in mainframe computers. The metrics that mattered to management and customers were cost per megabyte of storage, access time, and total capacity. The existing 14-inch drive won on all three: It was cheaper per megabyte and much faster and had roughly ten times the total storage capacity. The 8-inch drive could offer only smaller volume, lighter weight, and lower total cost. For these reasons, mainframe manufacturers, such as IBM, stayed with the 14-inch drive. Competing disk-drive manufacturers (e.g., Quantum Corporation), however, pursued the 8-inch drive and found ready customers at Wang Laboratories, Digital Equipment Corporation (DEC), and other makers of minicomputers. With

---

[101] See Clayton M. Christensen, *The Innovator's Dilemma: The Revolutionary Book That Will Change the Way You Do Business*, New York: HarpersCollins, 2006. For an application of this concept to innovation within the USN, see Terry Pierce, *Warfighting and Disruptive Innovation: Disguising Innovation*, London: Routledge, 2004. Christensen's work has recently been the subject of some high-visibility criticism that is, at least in part, a reaction to the business community turning the concept into a fad and so overusing and misapplying "disruptive innovation" that the concept has lost its original clarity and punch. See Justin Fox, "The Disruption Myth," *Atlantic*, September 17, 2014; referenced September 29, 2014; Jill Lepore, "The Disruption Machine: What the Gospel of Innovation Gets Wrong," *New Yorker*, June 23, 2014; referenced August 1, 2014; and Chris Newfield, "Christensen's Disruptive Innovation After the Lepore Critique," *Remaking the University*, June 22, 2014; referenced August 1, 2014. For Christensen's response to Lepore, see Drake Bennett, "Clayton Christensen Responds to *New Yorker* Takedown of 'Disruptive Innovation,'" *BloombergBusinessweek*, June 20, 2014; referenced August 1, 2014.

the 8-inch drive, Wang could offer customers affordable hard drives built into their minicomputers. The addition of a quality hard drive spurred demand for minicomputers, which, in turn, created a great demand for 8-inch drives. Within a few years, 8-inch drive performance had improved so much that it displaced the 14-inch drive in mainframes. IBM and other firms did try to jump into the 8-inch market but were so far behind Quantum and the others that they were never competitive. This was only the first in a series of disruptive changes in computer hard-disk technology—most notably, the 5.25-inch drive that made the personal computer possible and the 3.5-inch drive for laptop computers.[102] Christensen found similar disruptive processes in a wide range of industries, from communications to excavation.

Chinese improvements in ballistic missiles and supporting systems (mobile TELs, guidance from GPS or an inertial navigation system, remote sensing, over-the-horizon (OTH) radars,[103] submunition dispensers, and terminal guidance) have the potential to be disruptive in the military sphere in a manner that is similar to what Christensen describes in the commercial world.[104] With a force of more than 1,000 high-quality TBMs,[105] centralized in a single organization dedicated to missile operations (the 2nd Artillery Corps),[106] the Chinese field an offensive conventional missile force unlike any the world has seen. This force has the potential to conduct accurate and sustained attacks against U.S. and partner-nation airfields in East Asia. According to DoD,

> U.S. bases on Okinawa are in range of a growing number of Chinese MRBMs [medium-range ballistic missiles], and Guam could potentially be reached by air-launched cruise missiles. Chinese missiles have also become far more accurate and are now better suited to strike regional air bases, logistics facilities, and other ground-based infrastructure, which Chinese military analysts have concluded are vulnerabilities in modern warfare.[107]

---

[102] C. Christensen, 2006, pp. 3–22.

[103] With regard to China's reconnaissance strike capabilities, DoD notes that

> the PLA [People's Liberation Army] Navy also is improving its over-the-horizon (OTH) targeting capability with sky wave and surface wave OTH radars, which can be used in conjunction with reconnaissance satellites to locate targets at great distances from China (thereby supporting long-range precision strikes, including employment of ASBMs [antiship ballistic missiles]). (Office of the Secretary of Defense, *Annual Report to Congress: Military and Security Developments Involving the People's Republic of China 2014*, Washington, D.C., 2014, p. 40)

Also see Ian Easton, *China's Evolving Reconnaissance-Strike Capabilities: Implications for the U.S.–Japan Alliance*, Arlington, Va.: Project 2049 Institute, February 2014.

[104] We first explored this parallel in a 2008 study for the USAF. See Vick, 2011.

[105] DoD estimates that the Chinese have 1,100 SRBMs and have fielded an unspecified number of MRBMs and IRBMs. See Office of the Secretary of Defense, 2014, p. 6.

[106] For additional details on the 2nd Artillery Corps, see Mark A. Stokes and Ian Easton, *Evolving Aerospace Trends in the Asia–Pacific Region: Implications for Stability in the Taiwan Strait and Beyond*, Arlington, Va.: Project 2049 Institute, 2010.

[107] Office of the Secretary of Defense, 2014, p. 31.

Unlike attacks from aircraft or cruise missiles, which can be defeated through good air defenses,[108] ballistic-missile defenses are extremely expensive and have limited effectiveness, particularly against large attacks.[109]

Applying Christensen's disruptive innovation concept, the USAF, with its unmatched capability for high-end conventional warfare, can be thought of as the market-dominant firm. Rather than attempt to challenge this dominance in a symmetric way (e.g., by beating the USAF in air-to-air combat)—something the PLA had little hope of doing—the development of a large, high-quality conventional mobile missile force offered an indirect alternative that could defeat U.S. airpower before it left its land or sea base. Thus, U.S. market dominance in power projection, exemplified in the new American way of war described earlier, could be disrupted via the technology of the ballistic missile and a business model (strategy) that emphasized preemptive strikes against adversary bases.[110]

What makes ballistic missiles so disruptive to the American way of war? When tied to advanced sensors and a robust command and control system, mobile ballistic missiles offer some of the benefits of a first-class air force to nations whose air forces are weak or developing. Just as the USAF's ability to initiate military campaigns by penetrating defenses and delivering precision munitions against a wide range of targets has proven to be devastating to adversary conventional military forces, so might a quality mobile ballistic-missile force create similar effects, particularly if used preemptively. Although ballistic missiles are more costly than aircraft to deliver larger numbers of weapons during a prolonged conflict, they could be decisive in a short war under some conditions. Specifically, ballistic-missile attacks on air bases and aircraft carriers have the potential to destroy many aircraft and heavily damage runways and aircraft carriers,[111] disrupting sortie generation so severely that an adversary might achieve its military

---

[108] Which is not to say that the United States currently possesses a robust defense against aircraft and cruise missiles, just that it is a more manageable problem than ballistic-missile defense.

[109] The problem of intercepting ballistic missiles is complex and well beyond the scope of this report. For our purposes, the essence of the problem is that the United States has yet to deploy a theater missile defense system that can defeat large attacks and is unlikely to in the near future. The DoD 2010 review of ballistic-missile defense programs admits this, noting that "these capabilities exist in numbers that are only modest in view of the expanding regional missile threat." See DoD, *Ballistic Missile Defense Review Report*, Washington, D.C., February 2010, p. v. For an overview of the missile defense problem, see National Research Council, *Making Sense of Ballistic Missile Defense: An Assessment of Concepts and Systems for U.S. Boost-Phase Missile Defense in Comparison to Other Alternatives*, Washington, D.C.: National Academies Press, 2012.

[110] The parallel is helpful but imperfect. We are not suggesting that China will replace the United States as the market-dominant power projector, as can happen in business disruptive innovations but rather that Chinese innovations could overturn the U.S. model for power projection and thereby undermine or even end its dominance in conventional warfighting.

[111] DoD reports that the DF-21D MRBM "gives the PLA the capability to attack large ships, including aircraft carriers, in the western Pacific Ocean." See Office of the Secretary of Defense, 2014, p. 7. For more on Chinese antiship ballistic missiles, see Andrew S. Erickson and David D. Yang, "Using the Land to Control the Sea? Chinese Analysts Consider the Antiship Ballistic Missile," *Naval War College Review*, Vol. 62, No. 4, Autumn 2009, pp. 53–86. Some U.S. options to counter this threat are assessed in Thomas P. Ehrhard and Robert O. Work, *Range,*

and political objectives before the United States could effectively intervene.[112] This would present the United States with a fait accompli to either accept or attempt to reverse at potentially great cost. Additionally, adversary forces that possess ballistic missiles, cruise missiles, advanced aircraft, and SOF might be able to use them in mutually supporting ways to impair USAF and partner-nation capabilities.[113] Finally, missile mobility is a key attribute of these systems because, instead of presenting fixed targets at known locations (that can be attacked with standoff weapons), mobile missiles present an extraordinary search problem for the United States, requiring persistent high-resolution surveillance of a vast area and either rapid strike or the ability to track targets for hours while subsonic attack systems transit to the target location.[114]

This helps explain why mobile TBMs have been pursued with such determination by the PLA and, in retrospect, might even appear to be the obvious strategy. But from the perspective of U.S. airmen and defense planners watching these developments in the 1990s, the wisdom of the PLA missile investments seemed questionable. The unimpressive performance of Iraqi Scud missiles in 1991 reinforced the understanding that conventional ballistic missiles of that era were clearly inferior to manned aircraft. On the metrics that mattered to airmen,[115] not to mention programmers, planners, and strategists, ballistic missiles underperformed. The cost per target destroyed was much greater, accuracy much lower, payload per mission lower, range fixed (and too short), and force deployability and flexibility poor. If the Chinese wanted to waste money on missiles, that was their business; U.S. defense planners had no interest in such inflexible and brute-force weapons.[116] That slowly changed between 1999 and 2008, as a growing body of research, analysis, and war-gaming built a new consensus within the defense community that

---

Persistence, Stealth, and Networking: The Case for a Carrier-Based Unmanned Combat Air System, Washington, D.C.: Center for Strategic and Budgetary Assessments, 2008.

[112] The broader implications of this threat are explored in Christopher J. Bowie, *The Anti-Access Threat and Theater Air Bases*, Washington, D.C.: Center for Strategic and Budgetary Assessments, 2002; Vitaliy O. Pradun, "From Bottle Rockets to Lightning Bolts: China's Missile Revolution and PLA Strategy Against U.S. Military Intervention," *Naval War College Review*, Spring 2011, pp. 7–39; Ian Easton, *China's Military Strategy in the Asia– Pacific: Implications for Regional Stability*, Arlington, Va.: Project 2049 Institute, September 26, 2013; and Evan Braden Montgomery, "Contested Primacy in the Western Pacific: China's Rise and the Future of U.S. Power Projection," *International Security*, Vol. 38, No. 4, Spring 2014, pp. 115–149.

[113] For a discussion of joint 2nd Artillery Corps and PLA Air Force offensive operations, see Cliff, Fei, et al., 2011, pp. 184–186.

[114] The difficulties of finding mobile TELs in China are assessed in Alan J. Vick, Richard M. Moore, Bruce R. Pirnie, and John Stillion, *Aerospace Operations Against Elusive Ground Targets*, Santa Monica, Calif.: RAND Corporation, MR-1398-AF, 2001, pp. 57–81.

[115] Also the fact that the U.S. Army, not the USAF, developed and controlled SRBMs, MRBMs, and IRBMS likely lessened the attention that airmen paid to Chinese TBM development. Thanks to reviewer Thomas Mahnken for this insight.

[116] Defense planners did, however, recognize that ballistic and cruise missiles were a threat to U.S. forces and rear areas as early as 1993. For example, the October 1993 *Bottom Up Review* lists among the tasks to be accomplished in phase 1 of a major regional conflict: "Protect friendly forces and rear-area assets from attack by aircraft or cruise and ballistic missiles, using land- and sea-based aircraft, ground- and sea-based surface-to-air missiles." See Aspin, 1993, p. 16. Thanks to reviewer David Ochmanek for noting this.

Chinese ballistic missiles represented a growing threat to U.S. power-projection capabilities and that counters would be necessary.[117] For example, Col Jordan Thomas, until recently the senior USAF officer in the Pentagon's joint Air–Sea Battle Office, observed, "Twenty years ago [U.S. military personnel] were not under a ballistic missile threat—or at least not a credible ballistic missile threat. Today they are."[118]

The initial USAF and DoD reaction to the Chinese developments appears to be consistent with patterns identified in the scholarship on military innovation. In his analysis of foreign military innovation and U.S. intelligence from 1918 to 1941, Thomas Mahnken identified several barriers to recognizing foreign innovation:

> First, intelligence agencies are more inclined to monitor the development of established weapons than to search for new military systems. Second, intelligence agencies pay more attention to technology and doctrine that have been demonstrated in war. . . . Finally, it is easier to identify innovation in areas that one's own services are exploring than those they have not examined, are not interested in, or have rejected.[119]

Turning back to Christensen's concept, there are many similarities between it and military innovation: The dominant firm in each case has clearly defined metrics that accurately reflect the "value chain" of both the firm and its customers; the disruptive innovation is unattractive when tested on these metrics; the disrupter believes that there is a market for the innovation with customers who value other metrics; the innovation greatly improves in performance once in the field and being manufactured on a relatively large scale; and, finally, the innovation is disruptive to the dominant firm. All this has already happened with ballistic missiles. On the other hand, in the commercial world, the dominant firm needs to adopt the innovation if it wishes to remain competitive. In military innovation, that is not necessarily the case. The United States might benefit from the development of a conventional TBM force, but land- and sea-based aircraft and air-launched missiles are likely to remain a better fit (because of their versatility and ease of deployment) for a global power.[120] The challenge for the United States is, therefore, not

---

[117] The first analyst to argue in print that the PLA was seeking to develop highly accurate conventional ballistic missiles was Mark Stokes. See Mark A. Stokes, *China's Strategic Modernization: Implications for the United States*, Carlisle, Pa.: Strategic Studies Institute, U.S. Army War College, September 1999. See also Thomas Christensen, "Posing Problems Without Catching Up: China's Rise and Challenges for U.S. Security Policy," *International Security*, Vol. 25, No. 4, Spring 2001, pp. 5–40; David A. Shlapak, David T. Orletsky, and Barry Wilson, *Dire Strait? Military Aspects of the China–Taiwan Confrontation and Options for U.S. Policy*, Santa Monica, Calif.: RAND Corporation, MR-1217-SRF, 2000; and Cliff, Burles, et al., 2007.

[118] John A. Tirpak, "Fighting for Access," *Air Force Magazine*, July 2013, p. 23.

[119] Thomas G. Mahnken, *Uncovering Ways of War: U.S. Intelligence and Foreign Military Innovation, 1918–1941*, Ithaca, N.Y.: Cornell University Press, 2002, p. 4.

[120] For an argument in favor of the United States developing its own conventional IRBM force, see Jim Thomas, vice president and director of studies, Center for Strategic and Budgetary Assessments, "Statement Before the House Armed Services Subcommittee on Strategic Forces on the Future of the INF Treaty," July 17, 2014.

necessarily how to adopt the disruptive innovation but rather how to counter it in an affordable way, a topic that is addressed more fully in the next chapter.

Chapter Five. Defensive Options

As noted at the beginning of this report, attacks on air bases date back 100 years to the early weeks of World War I, when both British and German aircraft made minor attacks on enemy airfields (without causing any damage).[121] Since the first successful attack in October 1914 on the Zeppelin base at Dusseldorf, airmen have embraced a wide range of measures to defend air bases.[122] These have included camouflage, concealment, and deception (CCD); hardening of facilities; dispersal of aircraft (on airfields, away from airfields, and across multiple airfields); active defenses; and postattack recovery. Although the technologies have become much more sophisticated in the past century, all these defensive counters date back to World War I. In this chapter, we briefly discuss CCD, hardening, dispersal, and postattack recovery.[123]

Active defenses, which include air base ground defense forces;[124] counter–rocket, artillery, and mortar systems; anti-aircraft artillery (AAA); surface-to-air missiles; ballistic-missile defenses; early-warning and other radars; electronic warfare systems; fighter interceptors; and the command, control, and communication network to integrate them, are not discussed in this report. Active defense is a vast and technically complex topic that includes local-, area-, and theater-level defenses. It also involves difficult doctrinal and joint issues because either the USAF is underresourced to counter all threats (e.g., it relies on U.S. or partner ground forces to supplement USAF security forces when facing larger ground threats) or it has no relevant capability. For example, the USAF has no organic ground-based defenses against aircraft; armed remotely piloted vehicle;[125] or cruise-missile, ballistic-missile, rocket, artillery, or mortar attack, a problem noted by several USAF readers of a draft of this report. Air base defense against these threats is a vitally important topic but is beyond the scope of this analysis.

---

[121] Kreis, 1988, p. 6.

[122] Indeed, Dusseldorf had defenses in the form of medium machine guns emplaced on top of the Zeppelin shed. Although not able to prevent the attack, the defenders inflicted sufficient damage to the attacking aircraft that the pilot had to put the aircraft down in a farmer's field in Belgium and complete his return to base via bicycle. The aircraft was subsequently lost to German forces as their offensive closed on Antwerp. See Castle, 2011, pp. 22–26.

[123] The most thorough treatment of modern air base operability concepts and techniques is S. J. Sidoti, *Airbase Operability: A Study in Airbase Survivability and Post-Attack Recovery*, Canberra: Aerospace Centre, 2001.

[124] The most comprehensive treatment of emerging ground threats to air bases is Caudill's *Defending Air Bases in an Age of Insurgency*.

[125] See Caudill, 2014, pp. 306–307, for a discussion of remotely piloted vehicle threats.

Camouflage, Concealment, and Deception

CCD efforts seek to mislead the adversary about the locations of both airfields and aircraft.[126] Air base CCD dates back to World War I, when the Germans used camouflage paint to make aircraft harder to spot on the ground (or from above while in flight). Late in the war, the Germans were so disturbed by a successful attack on Boulay aerodrome (which destroyed five aircraft and damaged another ten) that they built a dummy airfield nearby. The decoy airfield succeeded, diverting subsequent British attacks from the real airfield.[127]

CCD efforts became much more sophisticated and common during World War II. A superb example of state-of-the-art CCD thinking in the United States in the early 1940s is found in *Modern Airfield Planning and Concealment*, a book written by Army Air Corps MAJ Merrill De Longe before Pearl Harbor but not cleared for publication until 1943. The book is a detailed analysis of how to design airfields to minimize visibility and vulnerability. It includes chapters (and many diagrams and photo illustrations) on concealment, camouflage, use of highways as airstrips, on-base dispersal of aircraft, and aircraft shelter options.[128]

Many of these ideas were put into practice during World War II. The British, in particular, elevated airfield deception measures to an art. The RAF conceived and executed an elaborate deception program built around two classes of decoy airfields. "Q sites" were intended to deceive night attackers. They were built in open areas near RAF bases (generally a mile or two from the base) and along typical flight paths where German aircrews would come upon the decoy prior to arriving at the real air base. Using elaborate lighting to mimic runway lights, moving aircraft and vehicles, they looked like RAF bases that had attempted light blackouts but without complete success. The imprecise navigation techniques of the era, combined with the natural desire of German aircrews to get the mission over, tricked many crews into attacking the decoy bases prior to reaching the actual airfield. As German bombs fell on the decoy bases, RAF personnel remotely triggered pyrotechnic devices that set fire to large piles of flammable debris, such as lumber waste and old tires, thus reinforcing the deception. Because German reconnaissance aircraft would typically overfly target areas the next day, the actual bases also implemented measures to complete the deception.[129] These included dragging damaged and destroyed aircraft onto ramps and placing canvas paintings of bomb craters on the operating surfaces. The fake craters came in two versions: one that displayed crisp shadows for sunny days and one with more-subdued tones for use on cloudy days. They were so realistic that many RAF pilots, when on approach to such an airfield, radioed the control tower seeking to abort the landing because of the craters. Finally, the RAF also constructed "K sites" designed to deceive

---

[126] See Sidoti, 2001, Chapter Eight, for a discussion of modern CCD techniques.

[127] Kries, 1988, p. 13.

[128] Merrill E. De Longe, *Modern Airfield Planning and Concealment*, New York: Chicago Pitman Publishing Corporation, 1943.

[129] The night decoy bases, made up of small lights in open fields, were essentially invisible during the day.

day attackers. These were much more elaborate than the night decoy sites and included runways, hangars, decoy aircraft, and vehicles—all infrastructure that would be expected at a real airfield. Both the day and night decoy airfields were remarkably effective and were attacked more often than the real ones (440 versus 430 attacks).[130]

Deception was used in the Pacific theater as well. For example, in preparation for major air attacks on Japanese airfields near Wewak, New Guinea, U.S. Fifth Air Force needed to create a forward base within fighter range to provide escort for its bombers. Tsili Tsili, New Guinea, was identified as the preferred location for the air base. As construction efforts went quietly forward there, Allied engineers made a much more visible effort to build a field closer to Wewak at Bena Bena, drawing both air and land attacks. While Japanese attention was focused on Bena Bena, the fighter field at Tsili Tsili was completed and a strong fighter force deployed there. The Japanese eventually discovered the real airfield but too late to prevent heavy air attacks on Wewak.[131]

All World War II combatants used CCD to some extent, with camouflage netting, concealment within tree lines, and decoy aircraft (typically aircraft wrecks rather than purpose-built decoys) the most widely used techniques. CCD efforts continued after World War II with extensive use of camouflage, as well as decoy aircraft, dummy craters, and fake airfields. Combatants used these techniques in (at minimum) the Korean, Vietnam, and 1971 India–Pakistan wars.[132] During the Cold War, NATO aircraft shelters were placed in tree lines where possible, and Warsaw Pact shelters were covered with soil or sod, offering partial concealment. Figure 5.1 illustrates how NATO Cold War–era air bases (in this case, Ramstein AB, Germany) placed fighter shelters on loops in woods. Although the shelters are clearly visible from directly overhead in this image, placing them in woods would have made target acquisition and attack more difficult for Warsaw Pact fighters. That is because, in the 1970s and 1980s, Warsaw Pact fighters would have had to visually spot the shelters and deliver dumb bombs from low or medium altitude, where tree lines would have somewhat obscured line of sight to shelters, particularly under the frequent cloudy, rainy, or low light (in winter) conditions of northern Europe.

---

[130] Seymour Reit, *Masquerade: The Amazing Camouflage Deceptions of World War II*, New York: Hawthorn Books, 1978, pp. 49–61.

[131] Bergerud, 2001, p. 629.

[132] See Kreis (1988) for Vietnam and Korea. Jagan Mohan and Chopra report that the Indian Air Force quickly realized that many pilot claims for aircraft destroyed on the ground at Pakistani airfields during the 1971 conflict were actually of dummy F-86 Sabres. See Jagan Mohan and Chopra, 2006, p. 151.

Figure 5.1. Hardened Aircraft Shelters on Loop, Ramstein Air Base, Germany

SOURCE: Google Earth.

In the late 1980s, the USAF developed inflatable decoy aircraft like the F-16 shown in Figure 5.2, but the Cold War ended before they were procured in numbers. Similar decoys are manufactured for various U.S. and Russian designs, but they do not appear to be in common use by major air forces.

Figure 5.2. F-16 Decoy and Real F-16

Photo courtesy of USAF. In this image, the decoy is nearest the camera, and the real F-16 is farther away.

Hardening

Hardening efforts seek to protect vital resources (e.g., aircraft, fuel, personnel, and command posts) from enemy attack by reducing the effective radius of attacking weapons or defeating them outright.[133] Airfield infrastructure was hardened somewhat during World War I. Initially, airfields were just collections of tents on open fields with aircraft parked in the open. After the front stabilized, air bases became more-permanent facilities. Hangars were built to protect aircraft from the elements and buildings erected for sleeping and eating quarters. Modest hardening was implemented, including dugouts built with log roofs, dirt and sandbags to protect personnel during air attack, and some use of sandbags and revetments to protect structures.[134]

---

[133] For a detailed discussion of modern hardening techniques, see Sidoti, 2001, Chapter Nine.

[134] Kreis, 1988, pp. 13, 16.

During World War II, military engineers developed a much greater capacity to build hardened structures. For example, the USN Mobile Construction Battalion undertook a crash effort to build underground tunnels for the protection of radio and radar equipment at Henderson Field, Guadalcanal. One of the most ambitious hardening programs of the war was the creation of a massive underground fuel-storage facility at Pearl Harbor that could protect 6 million barrels of fuel and diesel oil, complete with underground "bombproof" pump house. Begun in December 1940, the first phase was completed and ready for use by September 1942 and full project completed by July 1943.[135]

Hardening programs over the course of the following century included protective structures for aircraft, buried and hardened fuel and munitions, underground command posts, and other measures to make airfield infrastructure more resistant to attack. Base-hardening and recovery (e.g., runway repair) programs view the air base as a system whose primary purpose is the generation of aircraft sorties. These programs seek to identify critical vulnerabilities and mitigate effects through selective hardening and recovery capabilities. Much of the analytic work done at RAND and elsewhere on air base vulnerability has applied operations research techniques in order to better understand system dynamics and to identify the most cost-effective defensive counters. Because of the relative vulnerability and scarcity of aircraft, protecting them has received particular attention and is the focus of this section.

By World War II, attacking aircraft possessed the potential to do great damage to aircraft parked in the open, as demonstrated many times during that conflict, most notably during Operation Barbarossa in June 1941, the Japanese attacks on U.S. airfields on Oahu and in the Philippines in December 1941, and finally Operation Bodenplatte in January 1945. Large payloads, bigger guns and cannons, and more-accurate bombing techniques all made World War II fighter aircraft highly effective in strafing and dive-bombing attacks on airfields. As a result, three-sided aircraft revetments become the norm. These were most often constructed using soil, brick, and sandbags, often in combination, to create protective berms or walls. Revetments provided some protection from low-level strafing or bombing near misses, but, if the attacker made multiple strafing runs from different directions, losses could be heavy. Revetments also helped prevent secondary explosions and fires or munitions accidents from damaging nearby aircraft. Despite their limitations, revetments were used extensively by all combatants in World War II, Korea, Vietnam, the Arab–Israeli wars, ODS, and today by the USAF at Bagram AB, Afghanistan. It should be noted that, in some cases (e.g., Andersen Air Force Base on Guam during the Vietnam War), revetments were used to limit damage from accidental discharge of munitions rather than to protect from enemy attack.[136] Revetments were also common on NATO

---

[135] U.S. Bureau of Yards and Docks, *Building the Navy's Bases in World War II: History of the Bureau of Yards and Docks and the Civil Engineer Corps, 1940–1946*, Vol. II, Washington, D.C.: U.S. Government Printing Office, 1947, pp. 133–136, 244.

[136] Such accidents can be quite destructive. For example, one such munitions discharge at Bien Hoa AB, Vietnam in May 1965 destroyed 14 aircraft and damaged another 31. See R. Fox, 1979, p. 68.

and Warsaw Pact bases during the Cold War and are still in use at many airfields around the world. Figure 5.3 illustrates a simple World War II–era wall construction revetment with camouflage netting.

Figure 5.3. U.S. Navy TBD-1 Torpedo Bomber in Revetment, Naval Air Station Kaneohe Bay, Oahu, May 1942

SOURCE: U.S. Navy National Museum of Naval Aviation, 2014.

Figure 5.4 illustrates a larger and more robust brick and soil revetment protecting a Japanese G4M "Betty" bomber at a Japanese air base near Rabaul on the Pacific island of New Britain. This design was typical for Japanese air bases in the Pacific.

Figure 5.4. Japanese G4M Betty Bomber in Brick and Soil Revetment, Rabaul, New Britain, Papua New Guinea, 1943

SOURCE: Mortensen, 1950, p. 334.

To provide better protection from heavy or accurate attacks, aircraft needed to be parked either in caves or tunnels (quite effective but rarely available) or inside shelters constructed with hardened walls and roofs. The first aboveground hardened shelters were built by the USMC at Ewa Field, Oahu, beginning in September 1942 and are illustrated in Figure 5.5. Made of reinforced concrete and covered with soil and plants, the shelters offered excellent protection from anything other than a direct bomb hit or head-on strafing run.[137] Navy records document plans for 75 shelters, although a smaller number might have been built.[138] Because of the clamshell design, the width and length shrinks as one moves up the side of the shelter, and height drops as one moves toward the rear. Thus, this design would not be appropriate for larger aircraft with tricycle landing gear, but the design was fine for USMC tail-dragging aircraft, such as the Grumman F6F Hellcat, Grumman F4F Wildcat, and Vought F4U Corsair fighters. The effectiveness of the design, however, was not tested in combat. Japan never again attacked Ewa

---

[137] During a RAND research team visit on February 7, 2010, to the former site of Ewa Field, we measured and photographed some of the remaining shelters. There appeared to be about 30 still standing, many empty but others used for storage by local landowners. The shelters measured 18 ft. high and 55 ft. wide (floor level) in front and 39 ft. deep. Concrete facing was roughly 3 ft. high with 12 to 18 inches of concrete protection overall.

[138] The RAND team found shelters with identification numbers as high as 46, which suggests that close to 50 were built. For documentation of USN plans for 75 shelters, see U.S. Bureau of Yards and Docks, 1947, p. 144.

Field, and, once the island-hopping campaign began, the Marine Corps did not remain at any airfield under threat long enough to justify the effort and cost to construct hardened shelters.

Figure 5.5. U.S. Marine Corps Concrete Fighter Shelter, November 25, 1944, Ewa Field, Oahu

SOURCE: U.S. National Archives, U.S. Navy historical collection.

Modern hardened concrete shelters with doors first appeared during the Korean War. UN forces discovered two such shelters when they captured Wonsan Airfield in North Korea in October 1950.[139] The USAF explored shelter options during the 1950s, and, in 1962, U.S. Air Forces in Europe (USAFE) and PACAF identified requirements for 600 and 411 hardened Eglin fighter shelters, respectively.[140] The Eglin shelter was constructed out of steel with a concrete headwall and earth overburden. One was built at Eglin Air Force Base for testing, but the program was not pursued.

---

[139] Kries, 1988, p. 269.

[140] Benson, 1981, p. 114.

It was not until 1968 that the USAF, driven by an urgent wartime need, made shelter construction a priority. In response to VC and NVA mortar and rocket attacks that caused a tripling of aircraft losses (to more than 500 damaged or destroyed in 1968), the USAF embarked on the Concrete Sky program, a crash effort to build fighter shelters at all USAF MOBs in Vietnam. These shelters were made of corrugated steel with 18 inches of 4,000-psi reinforced concrete on top, designed to defeat mortars or rockets up to 122 mm. They had back walls but no front doors. Between 1968, and 1970, 373 shelters were constructed. This shelter design lived up to expectations, defeating rocket or mortar attacks in multiple documented cases.[141]

Overlapping with the Vietnam effort, in 1964–1965, the USAF conducted a major study called the Theater Air Base Vulnerability Evaluation Exercise (known as TAB VEE). This effort included operational simulations, design and construction of prototype shelters, and explosive tests. Implementation was delayed by the priority effort in Vietnam, but, by the spring of 1969, the TAB VEE program (by then a NATO effort) was building the first shelters at Bitburg, Ramstein, and Soesterberg air bases in Europe, using the initial Concrete Sky design but rapidly modifying it with doors and then moving from the TAB VEE to the second- and third-generation shelters found at USAF bases around the world. Like the earlier design, the second- and third-generation shelters had 18-inch concrete overhead protection but also added a back wall with exhaust vent and a 1-ft.-thick concrete door with an external metal latticework, as can be seen in Figure 5.6. This design could defeat small submunitions, as well as protect the aircraft from near misses from large unitary weapons. It was not intended to defeat precision attack. The main benefit of this design against emerging threats is that it forces would-be attackers to use either precision unitary weapons or larger submunitions. By the end of the Cold War, the United States had constructed roughly 1,000 such shelters in Europe and the Pacific.

_____

[141] R. Fox, 1979, p. 71. Also see Weitze, 2001, pp. 239–240.

Figure 5.6. U.S. Air Force F-22 and Hardened Aircraft Shelter at Kadena Air Base, Japan

SOURCE: Airman First Class Justin Veazie; U.S. Air Force photo.

Although using a different design with earth overburden, the Soviet Union and Warsaw Pact built hundreds of roughly similar fighter shelters intended to protect against unguided large unitary weapons or small submunitions. NATO allies, such as Germany and the Netherlands, also pursued their own designs, with protection varying from the U.S. standard to some shelters with up to 1 m of concrete overhead. The various U.S., Soviet, and European designs (as well as copies) can be found around the world in countries as diverse as Turkey, Djibouti, and Cuba. A multiyear RAND analysis of Google Earth imagery of more than 700 airfields found hardened aircraft shelters in 70 countries, in every region of the world. Figure 5.7 displays the countries possessing hardened aircraft shelters.

Figure 5.7. Nations Possessing Hardened Aircraft Shelters

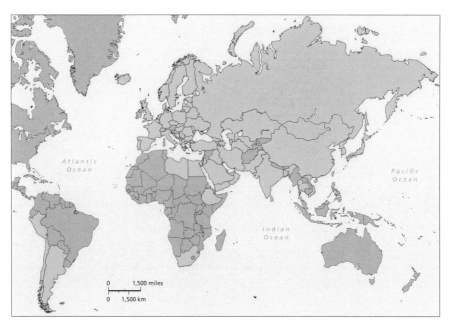

NOTE: The following countries have hardened aircraft shelters at one or more airfields on their territory:
Albania, Belgium, Bulgaria, Czech Republic, Denmark, Finland, France, Germany, Greece, Hungary, Iceland, Italy,
Netherlands, Norway, Poland, Portugal, Romania, Slovakia, Spain, Sweden, Switzerland, the United Kingdom,
Bahrain, Egypt, Iran, Iraq, Israel, Jordan, Kuwait, Lebanon, Libya, Oman, Qatar, Saudi Arabia, Turkey, United
Arab Emirates, Yemen, Azerbaijan, Belarus, Estonia, Georgia, Kazakhstan, Latvia, Lithuania, Moldova, Russia,
Turkmenistan, Ukraine, Uzbekistan, China, DPRK, Japan, Mongolia, Republic of Korea, Singapore, Taiwan,
Thailand, Vietnam, Bosnia and Herzegovina, Croatia, Kosovo, Macedonia, Serbia, Argentina, Chile, Cuba, Peru,
Ethiopia, Djibouti, South Africa, Bangladesh, India, and Pakistan.
RAND RR968-5.7

The most robust aboveground aircraft shelters are located in Saudi Arabia and Iraq. The Saudi design can be found at multiple bases, including Abha, Ta'if, and Dhahran. As Figure 5.8 illustrates, these drive-through shelters have massive overhead concrete protection and 2-ft.-thick steel doors protecting both entrances. Roofs have varying angles and are rubble-covered to hinder attack by penetrating weapons.

Figure 5.8. U.S. Air Force F-117 in Front of Shelter at Abha Air Base, Saudi Arabia, During Operation Desert Storm, 1991

Photo courtesy USAF History Office.

High berms parallel the shelters, creating alleys for taxiways on both sides and preventing low-angle weapon delivery against the doors. Finally, the ends of the shelter structures are blended into the ground to prevent weapon delivery against shelter walls. Iraqi shelters are freestanding individual structures that lack the protection of berms. They also have less overhead protection than Saudi shelters but nevertheless are massive concrete structures possessing much more protection than typical U.S., NATO, or Warsaw Pact designs. During ODS, the USAF found that 2,000 lb.–class bombs were usually required to defeat such shelters.

The final and most protective aircraft shelter concept would place aircraft underground either inside a cave or mountain tunnel or in an excavated area accessed by ramps or elevators. This idea dates back at least to 1934 when, in an interview or speech criticizing the recently released Baker Board report on aviation, Billy Mitchell reportedly "advocated underground airplane bases."[142] Underground shelters are also discussed in Arnold and Eaker's 1941 book *Winged*

---

[142] "Mitchell Calls It 'Whitewash,'" *New York Times*, July 23, 1934, p. 6.

Warfare, as well as De Longe's 1943 book on airfield planning.[143] Although one finds modern references to "underground" shelters, to our knowledge, the difficulties associated with moving aircraft up and down steep ramps or via elevators prevented his concept from ever being realized. Rather, the only shelters of this type in existence are the mountain tunnels accessed via a level taxiway that connects the shelter to a nearby runway.

Mountain shelters for aircraft can be found in Norway, Sweden, Switzerland, Taiwan, Kosovo, Croatia, and China and perhaps other locations.[144] Mountain shelters sometimes consist of single caverns with protective doors but more typically are tunnel complexes with multiple entrances. The aircraft inside the shelters are essentially invulnerable to attack, although they might be trapped inside the mountain if an attacker were able to strike the entrances, doors, and taxiways. For example, the DoD report on OAF describes attacks on entrances to an "underground aircraft storage and servicing facility" in Kosovo during OAF in 1999.[145] The feasibility of such attacks varies with local geology and entrance design, among other factors. Although mountain shelters offer a high degree of protection, they are an option at only a tiny fraction of potential operating locations and are costly and time-consuming to construct.

As the level of aircraft protection increases, at some point, aircraft are no longer an attractive target relative to other options. Aircraft in the open are clearly a high-leverage target. However, as the resources necessary to damage or destroy them in shelters increase, other targets, such as operating surfaces and fuel, become more attractive. Depending on the attacker's objectives, expected length of the conflict, and other considerations, closing air bases during critical phases of a conflict might be more important than destroying aircraft. For these reasons, the preferred mix and degree of hardening, postattack recovery, dispersal, and other air base defense measures can be determined only through modeling and simulation, particularly techniques that, like game theory, can account for the complexities and dynamics of opposed and adaptive strategies.

Protective structures for aircraft are likely to play an important role in future conflicts. For a few combatants who expect to fight from home territory and have the right geography, mountain shelters might be a major component of their air base defense strategy. Most combatants, however, will rely on some kind of aboveground shelter when necessary to protect aircraft, although the specific type of shelter design best suited for the 21st century is unclear. The thousands of legacy shelters at existing airfields offer good protection from ground threats, such as mortars, as well as from air attack by unguided unitary munitions or small submunitions.

---

[143] See Henry Harley Arnold and Ira Eaker, *Winged Warfare*, New York: Harper and Brothers, 1941, p. 71; and De Longe, 1943, Figure 58 and associated text on p. 96.

[144] This assessment is based on a combination of sources, including RAND analysis of Google Earth imagery of airfields, photos and text posted on the Internet by tourists who have visited former tunnel complexes (e.g., in Croatia and Sweden), and documentary video posted on the Internet (for Switzerland). An aviation museum is now housed in at least one former underground facility (at Goteborg, Sweden). See Aeroseum, home page, undated; referenced September 8, 2014.

[145] See DoD, *Report to Congress: Kosovo/Operation Allied Force After-Action Report*, Washington, D.C., January 31, 2000, p. 82.

These are likely to be used when available in future contingencies. That said, as precision threats proliferate, legacy shelters might not offer sufficient protection for all operational environments. Where land and funding are abundant, the Saudi-style shelters offer a high degree of protection from anything short of sustained attack with large precision penetrating weapons. For the United States, often operating in an expeditionary manner, the construction of such shelters (primarily on foreign soil) will, in most cases, be impractical for operational, economic, and political reasons. Permanent shelters also are immovable targets that simplify the adversary's targeting problem.[146] Protective structures of some type are likely to be a part of the future U.S. access strategy, but the level of protection, size, and number of new shelters will all have to be determined as part of an integrated strategy that includes all elements of air base defense (e.g., dispersion) and is developed in partnership with key host nations.

On-Base Dispersal

On-base dispersal of aircraft in the open and in wood lines (often combined with use of camouflage netting and decoy aircraft) is intended to make it more difficult to find and attack aircraft and to ensure that no single bomb can damage more than one aircraft.[147] Like most other air base resiliency techniques, it was first used extensively in World War II. An excellent example is found on the island of Malta, where the RAF dispersed aircraft among dozens of widely separated revetments and parking spots on its airfields and even had connecting taxiways between the Luqa and Hal Far airfields, allowing aircraft to use runways at either location.[148] In the Pacific theater, both the United States and Japan dispersed aircraft as widely as feasible given the more-difficult jungle and mountainous terrain common on islands, such as Guadalcanal and New Britain.

The Soviets' detonation of their first atomic bomb in 1949 meant that, within a few years, the Soviet Air Force could have the capacity to strike USAF and NATO air bases with nuclear weapons. Air base designs in the 1950s reflected this with aircraft dispersed in circular parking areas at the corners of bases. NATO also considered a range of dispersed basing options, ultimately settling on a combination of MOBs and dispersal bases (called colocated operating bases).[149] On-base dispersal was also used to some degree in Vietnam. For example, three weeks after the November 20, 1964, mortar attack on Bien Hoa AB, Gen Joseph Moore, 2nd Air

---

[146] Thanks to reviewer John Drew of RAND for this observation.

[147] See Sidoti, 2001, Chapter Ten, for a conceptual treatment.

[148] Kreis, 1988, p. 126.

[149] See Benson, 1981, p. 24, Figure 1, for an illustration of an on-base dispersal scheme that was implemented at NATO bases in France.

Division Commander, ordered dispersal pads to be constructed at Bien Hoa, Tan Son Nhut, and Da Nang.[150]

On-base dispersal has some potential to reduce losses from mortar or missile attack by limiting the damage that any single mortar round or missile can inflict. For this reason, it has been the subject of several recent studies.[151] Whether the benefits are sufficient to overcome the additional burdens placed on security forces, maintenance, and fuel personnel remains to be seen, but on-base dispersal is a natural complement to dispersing aircraft across many bases.

Dispersal Across Many Bases

Dispersing aircraft across many bases creates redundancy in operating surfaces and facilities. This enhances basic safety of flight by providing bases for weather or inflight-emergency diverts. It also increases the number of airfields that adversary forces must monitor and can greatly complicate their targeting problem (in part by raising the prospect that friendly forces might move among several bases). At the least, dispersal (because it increases the ratio of runways to aircraft) forces an attacker to devote considerably more resources to runway attacks than would be the case for a concentrated force.[152] It also greatly increases construction and operating costs to spread aircraft across many major bases. To mitigate these costs, dispersal bases tend to have more-modest facilities and, at times, might be nothing more than airstrips.

During World War I, the German air force built standby airfields and moved aircraft among them both to make it harder for British and French air forces to know where they were operating from and to reduce losses from attacks.[153] During World War II, all combatants dispersed aircraft across multiple landing fields to reduce base vulnerability. The Polish Air Force survived the initial, massive Luftwaffe attack against its airfields on September 1, 1939, because, in the 48 hours prior to the attack, all aircraft had been dispersed to emergency airfields.[154] The Soviet Air Force was the most dispersed and mobile of all World War II combatants, able to operate under the most austere conditions.[155] U.S. civil engineers became expert at rapidly building landing strips in South Pacific jungles, and both U.S. and Japanese air forces routinely operated from base clusters. For example, the Japanese operated from five airstrips in the vicinity of

---

[150] "Follow-Up to Bien Hoa Mortar Attack," 1965, p. 7.

[151] Marcus Weisgerber, "Pentagon Debates Policy to Strengthen, Disperse Bases," *Defense News*, April 13, 2014; referenced April 19, 2014.

[152] For an excellent conceptual treatment, see Sidoti, 2001, Chapter Ten.

[153] Kreis, 1988, p. 13.

[154] Halliday, 1987, p. 11.

[155] Benson, 1981, pp. 103–104

Rabaul, and U.S. forces on Guadalcanal operated aircraft from Henderson Field, as well as two auxiliary strips (Fighters 1 and 2).[156]

Dispersed operations were practiced, although only partially executed, by the USAF during the Korean War. Despite the limited capabilities of the DPRK air force, as late as November 1952, USAF Fifth Air Force leaders and senior staff remained concerned about the possibility of a large attack. For example, Col John Hearn, the Fifth Air Force director of intelligence, warned, "an initial, uninterrupted strike on the crowded airdromes at Kimpo and Suwon could destroy more than half of the F-86s . . . in Korea."[157] This led General Barcus, the Fifth Air Force commander, to order two squadrons of Sabres to the more remote Pusan airfield, despite the negative impact it would have on sortie generation. In January 1953, Barcus announced an ambitious dispersal plan named Doorstop, which created dispersal bases at Pusan, Taegu, Pohang, Pyongtaek, Kusan, and Osan-ni airfields, including emergency maintenance and sustainment stockpiles. Sabre squadrons routinely practiced deployments to these alternative airfields, a notable commitment to dispersal readiness given the daily wartime demands these units faced.[158]

During the Cold War, the Soviet Air Force embraced its World War II practice of highly dispersed operations. By 1957, Soviet and Warsaw Pact aircraft were dispersed among 218 primary and 536 secondary airfields through Eastern Europe and the western Soviet Union.[159] Similarly, USAFE sought to disperse aircraft across many locations.[160] By 1980, USAFE had access to 23 MOBs, five standby bases, and 72 colocated operating bases.[161] The Cold War ended with both NATO and the Warsaw Pact in possession of large base networks. Most of the NATO bases were closed during the 1990s, and USAFE now has only a handful of MOBs (Mildenhall and Lakenheath in the UK, Spangdahlem and Ramstein in Germany, and Aviano in Italy) and no formal system of dispersal. When the USAF went to war in 1991, it operated from many locations, but this was driven by the size of the deployment, not dispersal concerns. As discussed in Chapter Three, during the past two decades, the United States has operated in a highly efficient and effective manner that was optimized for that era of sanctuary.

The growing missile threat to USAF bases has led DoD to explore the full range of traditional air base defense measures, including dispersal across locations. For example, then–Deputy Assistant Secretary of Defense David Ochmanek noted in 2014, "Planners worry about what happens to our forward-based forces when they're inside the threat range from ballistic missiles and cruise missiles if those weapons are accurate and if they're deliverable in large

---

[156] See Bergerud, 2001.

[157] Quoted in Futrell, 1983, p. 661.

[158] Futrell, 1983, pp. 661–662.

[159] Benson, 1981, pp. 103–104.

[160] An excellent technical analysis of dispersal options in Europe toward the end of the Cold War is Halliday, 1987.

[161] Thanks to RAND colleague Stacie Pettyjohn, whose chart this is drawn from (original source, Benson, 1981).

numbers." Ochmanek also noted that analysis has shown "promising results" from "dispersing the force more radically" on and across U.S. air bases.[162]

A host of political, operational, and fiscal challenges must be overcome to move from a highly concentrated force posture to one that is widely dispersed. As NATO found when it abandoned its ambitious concept of the dispersed operating base in the 1950s,[163] developing and maintaining infrastructure at a large number of airfields can be prohibitively expensive. Dispersed operations also increase demands for security forces, distributed logistics, air and missile defense, and other support. In short, dispersed operations increase survivability of forces but lose economies of scale and efficiencies associated with more-concentrated operations. For these reasons, USAF MOBs will remain attractive for most peacetime activities and operations against lesser foes. Dispersal efforts are likely to be selective in investments and seek to leverage the large number of partner-nation military and civil airfields wherever possible. But even leveraging partner capabilities and investments, USAF support concepts and capabilities might need to change significantly to align with smaller force elements operating under attack and widely distributed across a potentially large area.

Postattack Recovery

Airfield repair concepts and capabilities evolved rapidly over the course of the 20th century. They were quite modest in World War I. For example, U.S. Army Air Service units were equipped with basic pioneer tools (shovels, axes, picks, and saws). Following an attack, craters in airfields or damage to simple structures could typically be repaired with local resources.

The greatest advances came during World War II, when state-of-the-art civil engineering techniques were integrated into military engineering units. This was particularly true in the U.S. military. Skilled civil engineers joined the war effort, bringing American experience in planning and managing major construction projects to air base construction. Bergerud argues that the U.S. dominance in this area had strategic effects in the South Pacific:

> In a conflict where air bases played a central role, the ability to construct and develop them was critical. Japan's rudimentary military construction capability proved a serious burden and crippled its forces during Guadalcanal. In contrast, the United States and its Allies showed an impressive ability to organize major construction projects both quickly and efficiently. Although no one guessed it at the time . . . the ability of the U.S. Command to change its priorities overnight and implement a major base construction plan in an area where no one had previously thought to fight eventually brought Japan's position in the South Pacific to ruin.[164]

---

[162] Weisgerber, 2014. See also PACAF, "U.S. Air Force B-52 Takes Part in Joint Training at RAAF Darwin," press release 010514, May 15, 2014; referenced June 10, 2014.

[163] See Benson, 1981, pp. 100–101, for details.

[164] Bergerud, 2001, p. 54.

The civil engineering capabilities resident in Army engineer and the new USN construction battalions (the Seabees) not only made it possible to rapidly construct main and auxiliary airstrips in support of island-hopping but also provided the foundation for airfield recovery efforts after attacks. Perhaps the most intense air attacks on a U.S. airfield in World War II were those experienced at Henderson Field on Guadalcanal in late 1942, when Japanese aircraft would attack during daylight and Japanese cruisers and battleships would shell the airfield during the night. These heavy attacks continued from mid-October until early November. If U.S. air forces (Army, USN, and USMC aircraft operating as an ad hoc joint force) had lost the ability to operate from Henderson, Guadalcanal most likely would have fallen to the Japanese. The aircraft were able to continue operating, if at times tenuously, because of the extraordinary efforts of maintenance personnel who rapidly repaired battle-damaged aircraft and Navy Seabees who did the same with operating surfaces. The official USN history of bases in World War II describes the action:

> On October 13, the enemy launched an all-out sea, air and land assault in an attempt to retake the island. About 30 twin-engined Japanese bombers dropped their bombs on the airfield, scoring several direct hits on the bomber-strip. U.S. fighter planes took off immediately in pursuit. As soon as the last plane left the ground, the entire battalion turned out to assist in repairing the damage. Special trucks, loaded with gravel to fill the bomb craters, had been standing by for just such an emergency. Others carried equipment for repairing the marston mat. . . . Entire sections were replaced and fitted into the undamaged mat.

> All the next day, the field was bombed again by enemy craft. . . . Holes were put in the strip as fast as they could be repaired, and then repaired as fast as they were made. Nevertheless, our aircraft were able to use the field throughout the attacks. . . . Within 48 hours the field had been hit 53 times.

> It is the proud record of the Seabees that . . . the field was never out of operation for more than four hours and in emergencies was always usable by fighter planes.[165]

The World War II Seabee experience captures the dynamic nature of contested air base operations. The U.S. experience in conflicts since then has not been particularly demanding on civil engineer and other postattack recovery capabilities. As discussed earlier, attacks on USAF bases in Korea were small and caused little damage. Attacks on USAF bases in Vietnam caused considerable damage to aircraft and, on a few occasions, produced spectacular fires and explosions when mortar or rocket rounds hit fuel or ammunition storage areas but only minor damage to operating surfaces and no known cases in which air operations were disrupted because of attacks. At most, attacks required removal of damaged or destroyed aircraft, clearing of debris on parking ramps, and (rarely) construction of new fuel or ammunition storage areas. During the Cold War, USAF postattack recovery capabilities (especially rapid runway repair) increased greatly at forward bases in Korea and Europe but were never tested in combat. Finally, during

---

[165] U.S. Bureau of Yards and Docks, 1947, pp. 244, 246.

OEF and OIF, mortar and rocket attacks on bases were common, but damage was so light that no significant postattack recovery capabilities were needed.

Although all the relevant offensive and defensive technologies have greatly advanced in the past 70 years, the intensity of effort required to keep an airfield open under attack in the 21st century is more likely to resemble the World War II experience than the relatively benign air base defense environments of Iraq and Afghanistan.

Toward More-Resilient Basing

Professional military institutions are designed to overcome the uncertainties and surprises associated with combat and, as a result, are inherently resilient in organization, capacity, and culture. That is, they are built to be robust and flexible so that they can absorb surprises, adapt, and continue to operate despite disruption to and degradation of their preferred procedures, processes, and local environment. This resiliency is obviously not absolute and can vary greatly even within a single service. Military services tend to have the greatest resiliency in their expected combat settings, with it declining as one moves from high-threat tactical settings toward more-secure rear areas. Security is a relative concept, but, at some point, in the rear (at least in most conflicts), there is an expectation of complete security from serious attack. Indeed, all modern military organizations depend on a vast rear-area infrastructure (industrial facilities, electrical power, communication, transportation networks, skilled workers) that is not designed to withstand serious attack. For example, a modern surface combatant, such as an *Arleigh Burke–* class destroyer, is resilient in its combat setting but depends on a highly vulnerable industrial facility (a rear-area port) to rearm and refit. As military technology and concepts change, the resiliency requirements for individual force elements might change significantly. Given the challenges to the American way of war described earlier in this report, it appears that the resiliency requirements for forward U.S. bases will grow in the next decade. In particular, air bases and forward air units will need a level of resiliency more typically associated with World War II than with more-recent operations, such as ODS.

Over the course of modern conflict, air bases have resided at various locations along this resiliency spectrum. Forward air bases from World War I through the Cold War typically expected to be attacked and embraced a range of measures that enhanced their ability to operate in the face of attack. These included postattack recovery and repair capabilities, as well as plans and procedures to operate from auxiliary locations during periods when primary facilities were unavailable for operations because of heavy damage.[166] More-fragile support activities were limited to rear areas, where the threat of attack was minimal or at least lower. More recently, U.S. and allied air bases have increasingly resembled advanced industrial facilities, designed to maximize output (i.e., sorties) rather than robustness in the face of attack.

---

[166] See Sidoti, 2001.

The conventional missile threat to air bases is now publicly acknowledged by USAF and DoD officials. For example, then–Secretary of the Air Force Michael B. Donley testified before Congress that the USAF was pursuing programs to make USAF bases "resilient in a number of threat scenarios," and Chief of Staff Gen Mark A. Welsh III added that hardened bases would be "mandatory" given the growing missile threat. Col Jordan Thomas (then director of the joint Air–Sea Battle Office) added that the USAF initiative would include additional hardened aircraft shelters and runway-repair capabilities, as well as "dispersal, concealment, and deception."[167]

These initiatives would ideally allow the USAF to absorb adversary attacks then rapidly reconstitute operations at current or backup locations. A capability to adapt to changed conditions without significant loss of efficiency is key. Consistently with lessons from World War II, resiliency capabilities include runway and airfield explosive ordnance disposal clearance and repair; robust and agile communications, logistics, and transportation systems; command procedures; and a culture that encourages initiative and tactical adaptation. Advances in rapid runway-repair techniques and materials are expected to greatly reduce the time necessary to return a damaged runway to operational status.[168]

Looking to the Future

This chapter briefly explored the primary classes of options available for air base defense: CCD; hardening of facilities; dispersal of aircraft (on airfields, away from airfields, and across multiple airfields); and postattack recovery. Along with air defenses, these techniques have been used in all major wars since World War I. Rarely is a single technique sufficient in isolation. Rather, airfield defenders typically use them all, with the relative balance among them tailored by the particular requirements of a conflict.

As U.S. defense planners look to future challenges in which anti-access threats are likely to be more common, they will need to develop a new portfolio of capabilities that includes concepts for more-dispersed and resilient operations; programs of record (e.g., for mobile command posts and communication, additional security forces, expanded logistics support, aircraft decoys, hardened shelters, and fuel-storage bladders); investment plans (e.g., to help partner nations improve dispersal-site infrastructure); organizational and manning changes (e.g., to allow USAF squadrons to operate from more-austere dispersed locations rather than MOBs); and, finally, a more integrated joint force approach to protecting forward locations.

---

[167] Tirpak, 2013, p. 25.

[168] See Emily A. Bradley, "Andersen Airmen Learn Innovative Airfield Damage Repair Capability," U.S. Air Force, February 28, 2014; referenced September 2, 2014; and U.S. Army Corps of Engineers, "Joint Project Develops Airfield Damage Repair Materials, Techniques," December 18, 2012; referenced September 2, 2014.

Chapter Six. Conclusions and Recommendations

After a quarter century in which the United States routinely conducted air operations from secure bases unmolested by enemy air or missile attacks, the USAF is now preparing for future conflict environments where air bases could come under heavy and sustained attack. As USAF leaders and planners consider how a more contested rear area will affect operations, force structure, and basing requirements, it is worth considering how air base attacks and defensive counters evolved in the past 100 years of air warfare.

This report offered a brief review of that history and described how it led in 1990 to a new American way of war. It then considered how emerging adversary capabilities are undermining that way of war and considered the ways in which defensive measures, from hardening to base recovery, can work together to defend against these new threats. Our research findings are presented in this chapter.

Findings

Air Base Attacks Have Been a Common Feature of Conflicts in the Past Century

Between 1914 and 2014, there were at least 26 conflicts in which air bases were attacked. The conflicts spanned the globe, including Central America, South America, Europe, Africa, the Middle East, Southwest Asia, South Asia, Southeast Asia, Australia, and Northeast Asia. Air bases have often been priority targets during the early phases of conflicts. Prominent examples include Germany's 1941 Operation Barbarossa against the Soviet Union, Japanese attacks on airfields on Oahu and in the Philippines in 1941, Israeli preemptive attacks against Arab airfields during the 1967 Six-Day War, and Indian and Pakistani air attacks at the start of their 1965 conflict. The United States has the most recent experience with airfield attacks, having struck adversary air bases in Iraq, Afghanistan, Serbia, and Libya in operations between 1990 and 2014. In World War II alone, Axis air forces lost more than 18,000 aircraft destroyed *on the ground* by U.S. Army Air Force, Navy, and Marine air attack. Air forces are not the only threat to air bases; ground forces have caused large losses to air forces in many conflicts. For example, the USAF lost roughly 1,600 aircraft damaged or destroyed on the ground by VC or North Vietnamese rocket and mortar attacks on U.S. bases in South Vietnam.

The Major Components of Air Base Defense First Identified in World War I Reflect Enduring Military Principles and Offer a Sound Framework for Air Base Defense Planning Today

From the earliest days of air combat, airmen recognized the threat to air bases and took steps to reduce vulnerabilities. For example, during the first successful attack on an airfield (the October 8, 1914, British attack on a German Zeppelin base), German defenders were prepared and employed medium machine guns against the (single) attacking aircraft. Although this failed to stop the attack, the British aircraft was damaged and had to land behind German lines. World War II saw great advances in defensive techniques, including hardened aircraft shelters, major deception efforts, wide dispersal of aircraft on and across bases, air base recovery, and the integration of radar, radio, observers, aircraft, and AAA into air defense networks. By World War II, air base defense had developed into a field of military study with its own set of guidelines, if not formal doctrine. For example, a portfolio approach to base defense that integrated all options into a defensive scheme was established as a best practice in the early 1940s and remains so in 2014. Thus, although the specific technologies continue to evolve, the basic outlines of sound air base defense are long established. Finally, the history of past airfield battles offers useful lessons for operators, planners, and policymakers today.

After the Cold War Ended, the United States Found That It Could Operate from Rear-Area Sanctuaries, and from This Flowed a New American Way of War

Shortly after the demise of the Soviet Union, the United States found itself in a conflict with Iraq, a nation that demonstrated substantial military capability. In response to the Iraqi invasion of Kuwait, the United States and allies, respecting Iraqi military potential, moved a massive air, ground, and naval force to the region. U.S. planners believed that superior U.S. air forces and ground-based air defenses could keep the Iraqi Air Force at bay and, as a result, based forces to maximize efficiency rather than minimize vulnerability to attack. Airfields were packed with aircraft and ports jammed with equipment and supplies. As it turned out, the Iraqis were unable to threaten these force concentrations in a serious way, and coalition air–ground forces routed the Iraqi Army in a handful of days (after it had been pummeled from the air for several weeks). This experience established a new American way of war that became the template for OAF, OEF, OIF, and Operation Odyssey Dawn. The new way of war did the following:

- Rapidly deploy large joint forces to forward bases and littoral seas.
- Create rear-area sanctuaries for U.S. forces through air superiority.
- Closely monitor enemy activities while denying the enemy the ability to do the same.
- Begin combat operations in the manner and at the time and place the United States chooses.
- Seize the initiative with a massive air and missile campaign focused on achieving air superiority in the opening hours or days.

- Maintain the offensive initiative through parallel and continuous air operations throughout the depth of the battlespace.
- Sustain the air campaign from sortie factories—air bases and carriers generating large numbers of aircraft sorties (ISR, strike, refueling) with industrial efficiency and uninterrupted by enemy action.

This way of war served the United States well in the past two decades. As is the case with any period of military dominance, however, new technologies and concepts eventually reduce imbalances between opposing force capabilities. This appears to be happening today because of the proliferation of long-range strike systems.

Emerging Long-Range Strike Capabilities Are Bringing the Era of Sanctuary to an End, with Significant Implications for the American Way of War

Highly accurate long-range strike systems, particularly ballistic and cruise missiles, make it much harder to have useful rear-area sanctuaries because these missiles are extremely difficult to defend against and can reach those locations to which U.S. forces would prefer to deploy or from which they would prefer to operate. These preferences are driven by simple geography. The United States relies primarily on relatively short-ranged fighter aircraft for most tactical air missions (e.g., close air support, interdiction, air superiority, and suppression of enemy air defenses) and, although fighters can sustain operations from as far as 1500 nm, bases within 500 miles of operating areas are greatly preferred. Similarly, ground forces have a limited ability to transit long distances over land and generally depend on major ports within 500 miles of their operating areas. Advances in man-portable precision strike systems, although a lesser threat, also raise the possibility that adversary SOF might be more lethal in the future than in past conflicts. For these reasons, against some adversaries, U.S. forces will deploy to bases that are under risk of attack from the first day. A period of buildup under conditions of sanctuary can no longer be taken for granted against the most capable adversaries.

The United States Will Need to Adapt Its Power-Projection Concepts to Operate Under a Greater Threat of Attack

Power-projection concepts will likely change in several ways. Without sanctuary, U.S. forces might need to deploy in a more deliberate and cautious manner, treating rear areas as combat zones and moving in a tactical rather than administrative fashion. Defensive forces and preparations, by necessity, will become a higher priority, likely delaying the movement of some offensive capabilities into theater. Additional resources will be needed to support CCD efforts that had not been necessary during the past several decades of operations. Deployments might be much less efficient as forces are moved in smaller waves through multiple, dispersed ports and airfields. If the enemy chooses to attack during the deployment phase, that could further slow down and disrupt force movements and might make it difficult for the United States to gain the initiative, at least in the early phases of combat. Prepositioning of logistics support equipment and changes to current support processes could improve the USAF's ability to deploy to dispersed locations and operate under attack.

As in the Past, a Combination of Measures Is Needed, but the Specific Mix Will Vary Depending on the Political Geography of the Region, Adversary Capabilities, and U.S. Objectives

Active defenses, hardening, CCD, dispersal, and resiliency all make unique contributions to the air base defense and operability. If resources were unlimited, defenders would want to fully avail themselves of every one at every base. In the real world, however, choices will have to be made. Extensive hardening, for example, might be a good choice at MOBs, where multiyear peacetime infrastructure projects are feasible. In contrast, hardening options are going to be much more limited at expeditionary locations because of time constraints if nothing else. In either case, hardening is not a panacea because it cannot be effective against all threats and the more-permanent structures make it possible for adversaries to do detailed peacetime targeting against these known locations. Similarly, although CCD possibilities should be considered at all locations, the best fit might be at dispersal or auxiliary bases, where hardening is more problematic, but CCD can exploit enemy preconceptions about how the United States will use such bases (e.g., how often and under what conditions forces will move among them). Modeling and simulation can identify the optimal mixes of each capability at individual bases and across the theater. In particular, it can clarify what options offer competitive advantages (i.e., force the enemy to spend more to counter than the measure costs the United States to implement). That said, in a major conflict, adversary actions will disrupt optimal resource allocations, and, in many cases, local base defenders will have to improvise. A good grasp of best practices and adaptations from previous wars will be valuable in those situations.

Recommendations

This research leads to three primary recommendations for USAF and DoD planners:

- **Consider the air base, the airspace above and near it, and the surrounding land as a battlespace, a place where defenders cannot expect sanctuary.** Too often, base defense and recovery are treated as support functions to be delegated to security forces and civil engineers. Although base and wing commanders take base defense seriously, it has not been a priority for the institutional air force, primarily because it has not been conceptualized as a core warfighting problem. It also has not received the attention and resources from the joint community, a critical problem because ground-based air defense of air bases is an Army responsibility. The relatively low priority for air base defense has led to a variety of shortfalls in USAF capabilities and in Army ground-based air defense capabilities. Understanding the air base as a battlespace makes its defense a core mission for the USAF and should help build a consensus among senior leaders to push forward new concepts, doctrine, and capabilities for operating under attack, possibly including the creation of USAF air and missile defense units.

- **Develop and test new concepts of operation for deployment to and operation of air bases under attack that incorporate both historical lessons and a full appreciation of emerging threats.** The American way of war as developed and executed between 1990 and 2011 is beginning to fray. Although current practices might remain viable for many years against weaker foes, it is time to begin developing alternative concepts and capabilities for more-contested environments in which disruptions to lines of communication, resource constraints, and local improvisation might be the rule rather than the exception. The U.S. experience during 1942 and early 1943 is particularly instructive because U.S. forces routinely operated under (and overcame) such conditions. New concepts can be developed and tested at relatively low cost using existing forces and infrastructure, as the USAF did in 2014 with its Rapid Raptor concept.[169] For example, tests could explore the practicality of dispersing aircraft at varying intervals across existing ramp space at a single base, of dispersing small packages of aircraft across multiple bases, or periodically moving units among a larger set of military and civilian airfields.
- **Explore organizational options to better support distributed and dispersed operations.** Although the mix of air base defense measures will vary by theater and threat, dispersed operations are likely to become central to operational concepts against highly capable adversaries. The full exploitation of dispersed concepts might, therefore, require changes to USAF organization and support structures. These are designed to maximize economies of scale by operating large forces from relatively few MOBs. An organization built around the air wing and intended to operate the wing at a single location (or at most two locations) might not be well suited for combat environments that require small force elements (e.g., squadron size or smaller) to be widely dispersed among many locations.

Final Thoughts

The USAF is no newcomer to the problem of air base defense. It experienced heavy attacks on air bases during World War II, lost 500 aircraft (damaged or destroyed) to mortar and rocket attacks during a single year of the Vietnam War, and expected to fight under heavy and sustained attack in the event of a war on Europe's Central Front. The USAF and other services overcame these wartime challenges, developing defensive measures that enabled effective air operations to continue despite ongoing attacks. Similarly, NATO air forces (with the USAF playing a leading role) developed a robust set of air base defense concepts and capabilities that helped sustain the credibility of NATO's conventional forces as a deterrent to Warsaw Pact aggression.

Although the emerging missile threat to air bases presents unique and difficult challenges, the USAF can meet and overcome them just as it did in earlier conflicts. The solution set will likely have both strong similarities and stark differences with previous air base defense concepts. We can, however, be sure of one similarity: Winning the battle of the airfields will require a level of

---

[169] Michael Trent Harrington, "Rapid Raptor Moves JBER F-22s Closer to the Fight," Joint Base Elmendorf-Richardson, May 15, 2014; referenced September 4, 2014.

institutional commitment to air base defense (from the USAF, DoD, and Congress) not seen since the height of the Cold War.

Bibliography

"120 mm MAT-120 Cargo Bomb," *Jane's Infantry Weapons*, updated October 11, 2011; referenced September 12, 2013.

Aeroseum, home page, undated; referenced September 8, 2014. As of January 29, 2015:
http://www.aeroseum.se/english/index.html

AeroVironment, "Switchblade," V0I.1, 2012; referenced May 26, 2014. As of January 29, 2015:
https://www.avinc.com/downloads/Switchblade_Datasheet_032712.pdf

"Army C-RAM Intercepts 100th Mortar Bomb in Iraq," *Defense Update*, July 2008; referenced September 4, 2014. As of January 29, 2015:
http://defense-update.com/newscast/0508/news/news2105_c_ram.htm

Arnold, Henry Harley, and Ira Eaker, *Winged Warfare*, New York: Harper and Brothers, 1941.

Aspin, Les, *Report on the Bottom-Up Review*, Washington, D.C.: U.S. Department of Defense, October 1993.

Association of the United States Army, "Army Prepositioned Stocks: Indispensable to America's Global Force-Projection Capability," Arlington, Va., December 2008; referenced April 3, 2014. As of January 29, 2015:
http://www.ausa.org/publications/torchbearercampaign/torchbearerissuepapers/documents/tb-ip_120308.pdf

Axford, Richard, and Marcus Gartside, "Complex Weapons in a Time of Austerity," presented at Royal Aeronautical Society Conference, June 12, 2012; referenced September 13, 2013.

Beevor, Antony, *The Battle for Spain: The Spanish Civil War 1936–1939*, London: Weidenfeld and Nicolson, 2006.

Bennett, Drake, "Clayton Christensen Responds to *New Yorker* Takedown of 'Disruptive Innovation,'" *BloombergBusinessweek*, June 20, 2014; referenced August 1, 2014. As of January 29, 2015:
http://www.businessweek.com/articles/2014-06-20/clayton-christensen-responds-to-new-yorker-takedown-of-disruptive-innovation

Benson, Lawrence R., *USAF Aircraft Basing in Europe, North Africa, and the Middle East, 1945–1980*, Ramstein Air Base, Germany: Headquarters, U.S. Air Forces in Europe, 1981; declassified 2011 by Air Force History Office.

Bergerud, Eric M., *Touched with Fire: The Land War in the South Pacific*, New York: Penguin Books, 1996.

————, *Fire in the Sky: The Air War in the South Pacific*, Boulder, Colo.: Westview, 2001.

Bergström, Christer, *Barbarossa: The Air Battle July–December 1941*, Hersham, Surrey: Midland/Ian Allen Press, 2007.

Biddle, Stephen, "Victory Misunderstood: What the Gulf War Tells Us About the Future of Conflict," *International Security*, Vol. 21, No. 2, Fall 1996, pp. 139–179.

Bonomo, James, Giacomo Bergamo, David R. Frelinger, John Gordon IV, and Brian A. Jackson, *Stealing the Sword: Limiting Terrorist Use of Advanced Conventional Weapons*, Santa Monica, Calif.: RAND Corporation, MG-510-DHS, 2007. As of January 29, 2015: http://www.rand.org/pubs/monographs/MG510.html

Bowie, Christopher J., *The Anti-Access Threat and Theater Air Bases*, Washington, D.C.: Center for Strategic and Budgetary Assessments, 2002.

————, "The Lessons of Salty Demo," *Air Force Magazine*, March 2009, pp. 54–57.

Bradley, Emily A., "Andersen Airmen Learn Innovative Airfield Damage Repair Capability," U.S. Air Force, February 28, 2014; referenced September 2, 2014. As of January 29, 2015: http://www.af.mil/News/ArticleDisplay/tabid/223/Article/473449/andersen-airmen-learn-innovative-airfield-damage-repair-capability.aspx

Calloway, Audra, "Picatinny Fields First Precision-Guided Mortars to Troops in Afghanistan," *www.army.mil*, March 29, 2011. As of January 29, 2015: http://www.army.mil/article/53988/

Campbell, Erin E., "The Soviet Spetsnaz Threat to NATO," *Airpower Journal*, Summer 1988; referenced June 9, 2014. As of January 29, 2015: http://www.airpower.maxwell.af.mil/airchronicles/apj/apj88/sum88/campbell.html

Castle, Ian, *The Zeppelin Base Raids: Germany 1914*, Oxford, UK: Osprey Publishing, 2011.

Caudill, Shannon W., *Defending Air Bases in an Age of Insurgency*, Maxwell Air Force Base, Ala.: Air University Press, 2014.

Chivers, C. J., "Qaddafi Troops Fire Cluster Bombs into Civilian Areas," *New York Times*, April 15, 2011. As of January 29, 2015: http://www.nytimes.com/2011/04/16/world/africa/16libya.html

Christensen, Clayton M., *The Innovator's Dilemma: The Revolutionary Book That Will Change the Way You Do Business*, New York: HarpersCollins, 2006.

Christensen, Thomas, "Posing Problems Without Catching Up: China's Rise and Challenges for U.S. Security Policy," *International Security*, Vol. 25, No. 4, Spring 2001, pp. 5–40.

Cliff, Roger, Mark Burles, Michael S. Chase, Derek Eaton, and Kevin L. Pollpeter, *Entering the Dragon's Lair: Chinese Antiaccess Strategies and Their Implications for the United States*,

Santa Monica, Calif.: RAND Corporation, MG-524-AF, 2007. As of January 29, 2015:
http://www.rand.org/pubs/monographs/MG524.html

Cliff, Roger, John F. Fei, Jeff Hagen, Elizabeth Hague, Eric Heginbotham, and John Stillion, *Shaking the Heavens and Splitting the Earth: Chinese Air Force Employment Concepts in the 21st Century*, Santa Monica, Calif.: RAND Corporation, MG-915-AF, 2011. As of January 29, 2015:
http://www.rand.org/pubs/monographs/MG915.html

Cornell, John T., *The Air Force in the Vietnam War*, Arlington, Va.: Aerospace Education Foundation, 2004.

Davis, Richard G., *The 31 Initiatives: A Study in Army–Air Force Cooperation*, Washington, D.C.: Office of Air Force History, U.S. Air Force, 1987. As of January 29, 2015:
http://purl.access.gpo.gov/GPO/LPS48159

De Longe, Merrill E., *Modern Airfield Planning and Concealment*, New York: Chicago Pitman Publishing Corporation, 1943.

Deptula, David A., *Effects-Based Operations: Change in the Nature of Warfare*, Arlington, Va.: Aerospace Education Foundation, 2001. As of January 29, 2015:
http://handle.dtic.mil/100.2/ADA466396

"DF-15," *Jane's Strategic Weapon Systems*, updated June 5, 2014; referenced September 8, 2014.

Diplomatic Conference for the Adoption of a Convention on Cluster Munitions, 108 signatories, August 1, 2010.

Director of Central Intelligence, *Warsaw Pact Nonnuclear Threat to NATO Airbases in Central Europe: National Intelligence Estimate*, Washington, D.C.: Central Intelligence Agency, NIE 11/20-6-84, October 25, 1984; declassified; referenced June 9, 2014. As of January 29, 2015:
http://www.foia.cia.gov/sites/default/files/document_conversions/89801/DOC_0000278545.pdf

DoD—*See* U.S. Department of Defense.

Don, Bruce W., Donald E. Lewis, Robert M. Paulson, and Willis H. Ware, *Survivability Issues and USAFE Policy*, Santa Monica, Calif.: RAND Corporation, N-2579-AF, 1988. As of January 29, 2015:
http://www.rand.org/pubs/notes/N2579.html

Dullum, Ove, *Cluster Weapons: Military Utility and Alternatives*, Oslo, Norway: Norwegian Defence Research Establishment, FFI-rapport 2007/02345, February 1, 2008. As of

January 29, 2015:
http://www.ffi.no/no/Rapporter/07-02345.pdf

Easton, Ian, *China's Military Strategy in the Asia–Pacific: Implications for Regional Stability*, Arlington, Va.: Project 2049 Institute, September 26, 2013. As of January 29, 2015:
http://www.project2049.net/documents/China_Military_Strategy_Easton.pdf

———, *China's Evolving Reconnaissance-Strike Capabilities: Implications for the U.S.–Japan Alliance*, Arlington, Va.: Project 2049 Institute, February 2014. As of January 29, 2015:
http://www.project2049.net/documents/
Chinas_Evolving_Reconnaissance_Strike_Capabilities_Easton.pdf

Ehrhard, Thomas P., and Robert O. Work, *Range, Persistence, Stealth, and Networking: The Case for a Carrier-Based Unmanned Combat Air System*, Washington, D.C.: Center for Strategic and Budgetary Assessments, 2008.

Emerson, Donald E., *AIDA: An Airbase Damage Assessment Model*, Santa Monica, Calif.: RAND Corporation, R-1872-PR, 1976. As of January 29, 2015:
http://www.rand.org/pubs/reports/R1872.html

———, *An Introduction to the TSAR Simulation Program: Model Features and Logic*, Santa Monica, Calif.: RAND Corporation, R-2584-AF, 1982. As of January 29, 2015:
http://www.rand.org/pubs/reports/R2584.html

Erickson, Andrew S., and David D. Yang, "Using the Land to Control the Sea? Chinese Analysts Consider the Antiship Ballistic Missile," *Naval War College Review*, Vol. 62, No. 4, Autumn 2009, pp. 53–86. As of January 29, 2015:
http://www.public.navy.mil/usff/sample/Pages/
Using-the-Land-to-Control-the-Sea--Chinese-Analyst.pdf

"Follow-Up to Bien Hoa Mortar Attack," Project CHECO staff report, Hickam Air Force Base, Hawaii: Headquarters Pacific Air Forces, December 1965; declassified by U.S. Air Force on January 9, 1991. As of January 29, 2015:
http://www.vietnam.ttu.edu/virtualarchive/items.php?item=F031100370386

Fox, Justin, "The Disruption Myth," *Atlantic*, September 17, 2014; referenced September 29, 2014. As of January 29, 2015:
http://www.theatlantic.com/magazine/archive/2014/10/the-disruption-myth/379348/

Fox, Roger, *Air Base Defense in the Republic of Vietnam, 1961–1973*, Washington, D.C.: Office of Air Force History, 1979.

Franks, Norman L. R., *Battle of the Airfields: Operation Bodenplatte, 1 January, 1945*, London: Grub Street, 1994.

Futrell, Robert Frank, *The United States Air Force in Korea: 1950–1953*, Washington, D.C.: Office of Air Force History, 1983.

Geneva International Centre for Humanitarian Demining, *A Guide to Cluster Munitions*, Geneva, November 2007.

Grissom, Adam, "The Future of Military Innovation Studies," *Journal of Strategic Studies*, Vol. 29, No. 5, 2006, pp. 905–934.

Hackett, John, *The Third World War: August 1985*, New York: MacMillan Publishing Company, 1979.

Hagen, Jeff, *Potential Effects of Chinese Aerospace Capabilities on U.S. Air Force Operations*, testimony presented before the U.S.–China Economic and Security Review Commission on May 20, 2010, Santa Monica, Calif.: RAND Corporation, CT-347, 2010. As of January 29, 2015:
http://www.rand.org/pubs/testimonies/CT347.html

Hall, Stephen C., "Air Base Survivability in Europe: Can USAFE Survive and Fight?" *Air University Review*, September–October 1982; referenced June 8, 2014. As of January 29, 2015:
http://www.airpower.maxwell.af.mil/airchronicles/aureview/1982/sep-oct/hall1.html

Halliday, John, *Tactical Dispersal of Fighter Aircraft: Risk, Uncertainty, and Policy Recommendations*, Santa Monica, Calif.: RAND Corporation, N-2443-AF, 1987. As of January 29, 2015:
http://www.rand.org/pubs/notes/N2443.html

Hallion, Richard, *Storm over Iraq: Air Power and the Gulf War*, Washington, D.C.: Smithsonian Institution Press, 1992.

Harrington, Michael Trent, "Rapid Raptor Moves JBER F-22s Closer to the Fight," Joint Base Elmendorf-Richardson, May 15, 2014; referenced September 4, 2014. As of January 29, 2015:
http://www.jber.af.mil/news/story.asp?id=123411157

Headquarters Military Assistance Command Vietnam, *Report of Investigation of Mortar Shelling of Bien Hoa Air Base on 1 November 1964*, Saigon, Vietnam, November 26, 1964; declassified by U.S. Air Force on January 9, 1991.

Higham, Robin, and Stephen John Harris, *Why Air Forces Fail: The Anatomy of Defeat*, Lexington, Ky.: University Press of Kentucky, 2006.

Hilburn, Matt, "'Shock Wave': Seabees Recount Deadly Mortar Attack," *Navy Times*, May 31, 2004.

Human Rights Watch, *Memorandum to CCW Delegates: A Global Overview of Explosive Submunitions*, prepared for the Convention on Conventional Weapons Group of Governmental Experts on the Explosive Remnants of War, May 21–24, 2002, Washington, D.C., 2002. As of January 29, 2015:
http://www.hrw.org/legacy/backgrounder/arms/submunitions.pdf

Jagan Mohan, P. V. S., and Samir Chopra, *The India–Pakistan Air War of 1965*, New Delhi: Manohar, 2006.

———, *Eagles over Bangladesh: The Indian Air Force in the 1971 Liberation War*, Noida, India: HarperCollins Publishers India, 2013.

James, T. C. G., *The Battle of Britain*, Portland, Ore.: F. Cass, 2000.

Kapuscinski, Ryszard, *The Soccer War*, New York: Vintage Books, 1992.

Keaney, Thomas A., and Eliot A. Cohen, *Gulf War Air Power Survey: Summary Report*, Washington, D.C.: U.S. Air Force, 1993.

Kirkland, Faris R., "The French Air Force in 1940: Was It Defeated by the Luftwaffe or by Politics?" *Air University Review*, September–October 1985; referenced September 29, 2014. As of January 29, 2015:
http://www.airpower.maxwell.af.mil/airchronicles/aureview/1985/sep-oct/kirkland.html

Kreis, John F., *Air Warfare and Air Base Air Defense, 1914–1973*, Washington, D.C.: Office of Air Force History, 1988.

Krepinevich, Andrew F., *Why AirSea Battle?* Washington, D.C.: Center for Strategic and Budgetary Assessments, 2010.

Krepinevich, Andrew F., Barry D. Watts, and Robert O. Work, *Meeting the Anti-Access and Area Denial Challenge*, Washington, D.C.: Center for Strategic and Budgetary Assessments, 2003.

Lambeth, Benjamin S., *The Winning of Air Supremacy in Operation Desert Storm*, Santa Monica, Calif.: RAND Corporation, P-7837, 1993. As of January 29, 2015:
http://www.rand.org/pubs/papers/P7837.html

———, *The Transformation of American Air Power*, Ithaca, N.Y.: Cornell University Press, 2000.

———, *NATO's Air War for Kosovo: A Strategic and Operational Assessment*, Santa Monica, Calif.: RAND Corporation, MR-1365-AF, 2001. As of January 29, 2015:
http://www.rand.org/pubs/monograph_reports/MR1365.html

———, *Air Power Against Terror: America's Conduct of Operation Enduring Freedom*, Santa Monica, Calif.: RAND Corporation, MG-166-1-CENTAF, 2006. As of January 29, 2015: http://www.rand.org/pubs/monographs/MG166-1.html

———, *The Unseen War: Allied Air Power and the Takedown of Saddam Hussein*, Annapolis, Md.: Naval Institute Press, 2013.

Lee, Richard R., *7AF Local Base Defense Operations, July 1965–December 1968*, Hickam Air Force Base, Hawaii: Headquarters Pacific Air Forces, July 1, 1969. As of January 29, 2015: http://www.dtic.mil/dtic/tr/fulltext/u2/a485052.pdf

Lepore, Jill, "The Disruption Machine: What the Gospel of Innovation Gets Wrong," *New Yorker*, June 23, 2014; referenced August 1, 2014. As of January 29, 2015: http://www.newyorker.com/magazine/2014/06/23/the-disruption-machine

"Lockheed Martin Missiles and Fire Control 227 mm Multiple Launch Rocket System (MLRS)," *Land Warfare Platforms: Artillery and Air Defence*, updated July 25, 2013; referenced May 26, 2014.

"M29 Cluster Bomb," National Museum of the U.S. Air Force, February 4, 2011; referenced May 26, 2014. As of January 29, 2015: http://www.nationalmuseum.af.mil/factsheets/factsheet.asp?id=15525

Mahnken, Thomas G., *Uncovering Ways of War: U.S. Intelligence and Foreign Military Innovation, 1918–1941*, Ithaca, N.Y.: Cornell University Press, 2002.

———, *Technology and the American Way of War Since 1945*, New York: Columbia University Press, 2008.

Manrho, John, and Ron Pütz, *Bodenplatte: The Luftwaffe's Last Hope*, Mechanicsburg, Pa.: Stackpole Books, 2010.

Mason, Tony, "Operation Allied Force: 1999," in John Andreas Olsen, ed., *A History of Air Warfare*, Washington, D.C.: Potomac Books, 2010, pp. 235–262.

McMillan Firearms, *TAC-50 McMillan Tactical Rifle*, date unknown; referenced May 26, 2014.

Michaels, Jim, and Charles Crain, "Insurgents Showing No Sign of Letting Up," *USA Today*, August 22, 2004; referenced June 9, 2014. As of January 29, 2015: http://usatoday30.usatoday.com/news/world/iraq/2004-08-22-iraq-cover_x.htm

Miller, Thomas Guy, *The Cactus Air Force*, New York: Harper and Row, 1969.

Milner, Joseph A., "The Defense of Joint Base Balad: An Analysis," in Shannon W. Caudill, ed., *Defending Air Bases in an Age of Insurgency*, Maxwell Air Force Base, Ala.: Air University, 2014, pp. 217–244. As of January 29, 2015: http://aupress.maxwell.af.mil/digital/pdf/book/b_0133_caudill_defending_air_bases.pdf

"Mitchell Calls It 'Whitewash,'" *New York Times*, July 23, 1934, p. 6.

Montgomery, Evan Braden, "Contested Primacy in the Western Pacific: China's Rise and the Future of U.S. Power Projection," *International Security*, Vol. 38, No. 4, Spring 2014, pp. 115–149.

Mortensen, Bernhardt L., "Rabaul and Cape Gloucester," in Wesley Frank Craven and James Lea Cate, eds., *Army Air Forces in World War II*, Vol. IV: *The Pacific—Guadalcanal to Saipan August 1942 to July 1944*, 1950, pp. 311–358. As of February 3, 2015: http://www.ibiblio.org/hyperwar/AAF/IV/AAF-IV-10.html

National Research Council, *Making Sense of Ballistic Missile Defense: An Assessment of Concepts and Systems for U.S. Boost-Phase Missile Defense in Comparison to Other Alternatives*, Washington, D.C.: National Academies Press, 2012. As of January 29, 2015: http://www.nap.edu/catalog.php?record_id=13189

Neu, C. Richard, *Attacking Hardened Air Bases (AHAB): A Decision Analysis Tool for the Tactical Commander*, Santa Monica, Calif.: RAND Corporation, R-1422-PR, 1974. As of January 29, 2015: http://www.rand.org/pubs/reports/R1422.html

Newfield, Chris, "Christensen's Disruptive Innovation After the Lepore Critique," *Remaking the University*, June 22, 2014; referenced August 1, 2014. As of January 29, 2015: http://utotherescue.blogspot.com/2014/06/christensens-disruptive-innovation.html

Office of Naval Intelligence, Air Branch, *Naval Aviation Combat Statistics: World War II*, Washington, D.C.: Naval Aviation History Office, June 17, 1946.

Office of the Secretary of Defense, *Annual Report to Congress: Military and Security Developments Involving the People's Republic of China 2013*, Washington, D.C., 2013. As of January 29, 2015: http://oai.dtic.mil/oai/oai?&verb=getRecord&metadataPrefix=html&identifier=ADA579650

———, *Annual Report to Congress: Military and Security Developments Involving the People's Republic of China 2014*, Washington, D.C., 2014. As of January 29, 2015: http://www.defense.gov/pubs/2014_DoD_China_Report.pdf

Oppel, Richard A., Jr., and Graham Bowley, "Rocket Fire Damages Plane Used by Joint Chiefs Chairman," *New York Times*, August 21, 2012. As of January 29, 2015: http://www.nytimes.com/2012/08/22/world/asia/top-us-commanders-plane-damaged-in-afghan-attack.html

O'Rourke, Ronald, *Cruise Missile Inventories and NATO Attacks on Yugoslavia: Background Information*, Washington, D.C.: Congressional Research Service, April 20, 1999. As of

January 29, 2015:
http://oai.dtic.mil/oai/oai?&verb=getRecord&metadataPrefix=html&identifier=ADA478026

Owen, Robert C., ed., *Deliberate Force: A Case Study in Effective Air Campaigning—Final Report of the Air University Balkans Air Campaign Study*, Maxwell Air Force Base, Ala.: Air University Press, 2000.

PACAF—*See* Pacific Air Forces.

Pacific Air Forces, "U.S. Air Force B-52 Takes Part in Joint Training at RAAF Darwin," press release 010514, May 15, 2014; referenced June 10, 2014. As of January 29, 2015:
http://www.pacaf.af.mil/news/story.asp?id=123411119

Peterson, Nolan, "Threat Remains for U.S. Troops in Afghanistan," United Press International, December 13, 2013; referenced December 16, 2013. As of January 29, 2015:
http://www.upi.com/Top_News/World-News/2013/12/13/
Threat-remains-for-US-troops-in-Afghanistan/8091386994110/

Pettyjohn, Stacie L., *U.S. Global Defense Posture: 1783–2011*, Santa Monica, Calif.: RAND Corporation, MG-1244-AF, 2012. As of January 29, 2015:
http://www.rand.org/pubs/monographs/MG1244.html

Pettyjohn, Stacie L., and Alan J. Vick, *The Posture Triangle: A New Framework for U.S. Air Force Global Presence*, Santa Monica, Calif.: RAND Corporation, RR-402-AF, 2013. As of January 29, 2015:
http://www.rand.org/pubs/research_reports/RR402.html

Pierce, Terry, *Warfighting and Disruptive Innovation: Disguising Innovation*, London: Routledge, 2004.

Plaster, John L., *The Ultimate Sniper: An Advanced Training Manual for Military and Police Snipers*, Boulder, Colo.: Paladin Press, 1993.

Pradun, Vitaliy O., "From Bottle Rockets to Lightning Bolts: China's Missile Revolution and PLA Strategy Against U.S. Military Intervention," *Naval War College Review*, Spring 2011, pp. 7–39. As of January 29, 2015:
https://www.usnwc.edu/getattachment/23a01071-5dac-433a-8452-09c542163ae8/
From-Bottle-Rockets-to-Lightning-Bolts--China-s-Mi

Press, Daryl G., "The Myth of Air Power in the Persian Gulf War and the Future of Warfare," *International Security*, Vol. 26, No. 2, Fall 2001, pp. 5–44.

"R-11/-17 (SS-1 'Scud'/8A61/8K11/8K14, and R-11FM [SS-N-1B])," *Jane's Strategic Weapon Systems*, updated January 7, 2014; referenced April 21, 2014.

Rafael Advanced Defense Systems, "Spike NLOS™: Multi-Purpose, Multi-Platform Electro Optical Missile," undated; referenced May 26, 2014. As of January 29, 2015:
http://www.rafael.co.il/marketing/SIP_STORAGE/FILES/6/1026.pdf

Rayment, Sean, "Harrier Destroyed by Afghan Rocket," *Telegraph*, October 16, 2005. As of January 29, 2015:
http://www.telegraph.co.uk/news/uknews/1500702/Harrier-destroyed-by-Afghan-rocket.html

Reit, Seymour, *Masquerade: The Amazing Camouflage Deceptions of World War II*, New York: Hawthorn Books, 1978.

"RGM/UGM-109 Tomahawk," *Jane's Strategic Weapon Systems*, updated May 6, 2014; referenced May 26, 2014.

Rodewig, Cheryl, "Geotagging Poses Security Risks," *www.army.mil*, March 7, 2012. As of January 29, 2015:
http://www.army.mil/article/75165/Geotagging_poses_security_risks/

Rostker, Bernard, *Information Paper: Iraq's SCUD Ballistic Missiles*, Washington, D.C.: U.S. Department of Defense, interim paper, July 25, 2000.

Rubin, Alissa J., "Audacious Raid on NATO Base Shows Taliban's Reach," *New York Times*, September 16, 2012. As of January 29, 2015:
http://www.nytimes.com/2012/09/17/world/asia/green-on-blue-attacks-in-afghanistan-continue.html

Sams, Kenneth, *Historical Background to Viet Cong Mortar Attack on Bien Hoa: 1 November 1964*, Project CHECO (Contemporary Historical Examination of Current Operations) Office, Headquarters 2nd Air Division, November 9, 1964; declassified by U.S. Air Force on January 9, 1991.

Schanz, Marc V., "Rethinking Air Dominance," *Air Force Magazine*, July 2013.

Secretary of the Air Force, *Facility Requirements*, Washington, D.C., Air Force Handbook 32-1084, September 1, 1996.

Shape Strategic Defense, home page, date unknown; referenced September 8, 2004.

Sherlock, Ruth, "Isil Fighters Capture Key Syrian Air Base in Sign of Growing Strength," *Telegraph*, August 24, 2014; referenced September 1, 2014. As of January 29, 2015:
http://www.telegraph.co.uk/news/worldnews/middleeast/syria/11054264/Isil-fighters-capture-key-Syrian-air-base-in-sign-of-growing-strength.html

Shlapak, David A., David T. Orletsky, Toy I. Reid, Murray Scot Tanner, and Barry Wilson, *A Question of Balance: Political Context and Military Aspects of the China–Taiwan Dispute*,

Santa Monica, Calif.: RAND Corporation, MG-888-SRF, 2009. As of January 29, 2015:
http://www.rand.org/pubs/monographs/MG888.html

Shlapak, David A., David T. Orletsky, and Barry Wilson, *Dire Strait? Military Aspects of the China–Taiwan Confrontation and Options for U.S. Policy*, Santa Monica, Calif.: RAND Corporation, MR-1217-SRF, 2000. As of January 29, 2015:
http://www.rand.org/pubs/monograph_reports/MR1217.html

Shlapak, David A., John Stillion, Olga Oliker, and Tanya Charlick-Paley, *A Global Access Strategy for the U.S. Air Force*, Santa Monica, Calif.: RAND Corporation, MR-1216-AF, 2002. As of January 29, 2015:
http://www.rand.org/pubs/monograph_reports/MR1216.html

Shlapak, David A., and Alan J. Vick, *"Check Six Begins on the Ground": Responding to the Evolving Ground Threat to U.S. Air Force Bases*, Santa Monica, Calif.: RAND Corporation, MR-606-AF, 1995. As of January 29, 2015:
http://www.rand.org/pubs/monograph_reports/MR606.html

Sidoti, S. J., *Airbase Operability: A Study in Airbase Survivability and Post-Attack Recovery*, Canberra: Aerospace Centre, 2001.

Sieff, Martin, "Analysis: Mortar Attack Fits Deadly Pattern," United Press International, January 7, 2004; referenced June 9, 2014. As of January 29, 2015:
http://www.upi.com/Business_News/Security-Industry/2004/01/07/AnalysisMortar-attack-fits-deadly-pattern/UPI-56791073525015/

Silver, Michael A., *Theater Airlifter Survivability on the Ground*, Wright-Patterson Air Force Base, Ohio: Air Force Institute of Technology, master's thesis, 1993. As of January 29, 2015:
http://oai.dtic.mil/oai/oai?&verb=getRecord&metadataPrefix=html&identifier=ADA262393

Stillion, John, "Fighting Under Missile Attack," *Air Force Magazine*, August 2009, pp. 34–37.

Stillion, John, and David T. Orletsky, *Airbase Vulnerability to Conventional Cruise-Missile and Ballistic-Missile Attacks: Technology, Scenarios, and U.S. Air Force Responses*, Santa Monica, Calif.: RAND Corporation, MR-1028-AF, 1999. As of January 29, 2015:
http://www.rand.org/pubs/monograph_reports/MR1028.html

Stokes, Mark A., *China's Strategic Modernization: Implications for the United States*, Carlisle, Pa.: Strategic Studies Institute, U.S. Army War College, September 1999. As of January 29, 2015:
http://purl.access.gpo.gov/GPO/LPS12109

Stokes, Mark A., and Ian Easton, *Evolving Aerospace Trends in the Asia–Pacific Region: Implications for Stability in the Taiwan Strait and Beyond*, Arlington, Va.: Project 2049

Institute, 2010. As of January 29, 2015:
http://project2049.net/documents/aerospace_trends_asia_pacific_region_stokes_easton.pdf

Tangredi, Sam J., *Anti-Access Warfare: Countering A2/AD Strategies*, Annapolis, Md.: Naval Institute Press, 2013.

Thomas, Jim, vice president and director of studies, Center for Strategic and Budgetary Assessments, "Statement Before the House Armed Services Subcommittee on Strategic Forces on the Future of the INF Treaty," July 17, 2014. As of February 3, 2015:
http://docs.house.gov/meetings/AS/AS29/20140717/102474/
HHRG-113-AS29-Wstate-ThomasJ-20140717-U1.pdf

Thompson, Wayne, *To Hanoi and Back: The United States Air Force and North Vietnam, 1966–1973*, Washington, D.C.: Smithsonian Institution Press, 2000.

Tirpak, John A., "Bombers over Libya," *Air Force Magazine*, July 2011, pp. 36–39.

———, "Fighting for Access," *Air Force Magazine*, July 2013.

Tol, Jan van, Mark Alan Gunzinger, Andrew F. Krepinevich, and Jim Thomas, *AirSea Battle: A Point-of-Departure Operational Concept*, Washington, D.C.: Center for Strategic and Budgetary Assessments, 2010.

Torgerson, Frederick, *Parked Aircraft Vulnerability to Mortar Attack*, Hickam Air Force Base, Hawaii: Headquarters Pacific Air Forces, September 1964; declassified by U.S. Air Force on August 23, 1990.

Trohanowsky, Raymond, U.S. Army Research, Development and Engineering Command Armament Research, Development and Engineering Center, "120mm Mortar System Accuracy Analysis," presented to International Infantry and Joint Services Small Arms Systems Annual Symposium, Exhibition, and Firing Demonstration, May 17, 2005. As of January 30, 2015:
http://www.dtic.mil/ndia/2005smallarms/tuesday/trohanowsky.pdf

USAF—*See* U.S. Air Force.

U.S. Air Force, *United States Air Force Statistical Digest World War II*, Washington, D.C.: Management Information Division, Directorate of Management Analysis, Comptroller of the Air Force, Headquarters, U.S. Air Force, December 1945; referenced September 24, 2013. As of January 29, 2015:
http://www.dtic.mil/dtic/tr/fulltext/u2/a542518.pdf

U.S. Army Corps of Engineers, "Joint Project Develops Airfield Damage Repair Materials, Techniques," December 18, 2012; referenced September 2, 2014. As of January 29, 2015:
http://www.erdc.usace.army.mil/Media/NewsStories/tabid/9219/Article/476446/
joint-project-develops-airfield-damage-repair-materials-techniques.aspx

U.S. Bureau of Yards and Docks, *Building the Navy's Bases in World War II: History of the Bureau of Yards and Docks and the Civil Engineer Corps, 1940–1946*, Vol. II, Washington, D.C.: U.S. Government Printing Office, 1947.

U.S. Department of Defense, *Conduct of the Persian Gulf War: Final Report to Congress*, Washington, D.C., April 1992.

———, *Report to Congress: Kosovo/Operation Allied Force After-Action Report*, Washington, D.C., January 31, 2000. As of January 29, 2015:
http://purl.access.gpo.gov/GPO/LPS16504

———, *Iraq's Scud Ballistic Missiles: Information Paper*, Washington, D.C., 2001.

———, *Ballistic Missile Defense Review Report*, Washington, D.C., February 2010. As of January 29, 2015:
http://handle.dtic.mil/100.2/ADA514210

———, "DoD News Briefing with Vice Adm. Gortney from the Pentagon on Libya Operation Odyssey Dawn," news transcript, March 19, 2011; referenced September 3, 2014. As of January 29, 2015:
http://www.defense.gov/transcripts/transcript.aspx?transcriptid=4786

U.S. Military Sealift Command, "Strategic Sealift (PM3)," undated; referenced April 4, 2014. As of January 29, 2015:
http://www.msc.navy.mil/PM3/

U.S. National Air and Space Intelligence Center, *Ballistic and Cruise Missile Threat*, Wright-Patterson Air Force Base, Ohio, 2013; referenced April 19, 2014. As of January 29, 2015:
http://oai.dtic.mil/oai/oai?&verb=getRecord&metadataPrefix=html&identifier=ADA582843

U.S. Navy National Museum of Naval Aviation, "File:TBD VT-3 at NAS Kaneohe Bay May 1942.jpeg," last modified November 12, 2014. As of February 3, 2015:
http://commons.wikimedia.org/wiki/
File:TBD_VT-3_at_NAS_Kaneohe_Bay_May_1942.jpeg

Vick, Alan J., *Snakes in the Eagle's Nest: A History of Ground Attacks on Air Bases*, Santa Monica, Calif.: RAND Corporation, MR-553-AF, 1995. As of January 29, 2015:
http://www.rand.org/pubs/monograph_reports/MR553.html

———, *Challenges to the American Way of War*, presentation to the Global Warfare Symposium, Los Angeles, Calif., November 17, 2011. As of January 29, 2015:
http://secure.afa.org/events/natlsymp/2011/scripts/AFA-111117-Vick.pdf

Vick, Alan J., Richard M. Moore, Bruce R. Pirnie, and John Stillion, *Aerospace Operations Against Elusive Ground Targets*, Santa Monica, Calif.: RAND Corporation, MR-1398-AF,

2001. As of January 29, 2015:
http://www.rand.org/pubs/monograph_reports/MR1398.html

Vish, Jeffrey A., *Guided Standoff Weapons: A Threat to Expeditionary Air Power*, Monterey, Calif.: Naval Postgraduate School, master's thesis, 2006. As of January 29, 2015: http://edocs.nps.edu/npspubs/scholarly/theses/2006/Sep/06Sep_Vish.pdf

Wald, Matthew L., "Citing Danger to Planes, Group Seeks Ban on a Sniper Rifle," *New York Times*, January 31, 2003; referenced June 9, 2014. As of January 29, 2015: http://www.nytimes.com/2003/01/31/national/31RIFL.html

Weigley, Russell Frank, *The American Way of War: A History of United States Military Strategy and Policy*, Bloomington, Ind.: Indiana University Press, 1977.

Weisgerber, Marcus, "Pentagon Debates Policy to Strengthen, Disperse Bases," *Defense News*, April 13, 2014; referenced April 19, 2014. As of January 29, 2015: http://www.defensenews.com/article/20140413/DEFREG02/304130017/ Pentagon-Debates-Policy-Strengthen-Disperse-Bases

Weitze, Karen, *Eglin Air Force Base, 1931–1991: Installation Buildup for Research, Test, Evaluation and Training*, Eglin Air Force Base, Fla.: Air Force Materiel Command, 2001.

Wohlstetter, Albert, Fred Hoffman, R. J. Lutz, and Henry S. Rowen, *Selection and Use of Strategic Air Bases*, Santa Monica, Calif.: RAND Corporation, R-266, 1954. As of January 29, 2015: http://www.rand.org/pubs/reports/R0266.html

Woods, Kevin M., *Iraqi Perspectives Project Phase II: Um Al-Ma'arik (The Mother of All Battles)—Operational and Strategic Insights from an Iraqi Perspective*, Vol. 1, Alexandria, Va.: Institute for Defense Analysis, P-4217, May 2008. As of January 29, 2015: http://handle.dtic.mil/100.2/ADA484530

Zaloga, Steve, *Poland 1939: The Birth of Blitzkrieg*, Oxford, UK: Osprey, 2002.